Lady (Mary Anne) Barker

First Lessons in the Principles of Cooking

In 3 Parts

Lady (Mary Anne) Barker

First Lessons in the Principles of Cooking
In 3 Parts

ISBN/EAN: 9783744749756

Printed in Europe, USA, Canada, Australia, Japan

Cover: Foto ©berggeist007 / pixelio.de

More available books at **www.hansebooks.com**

FIRST LESSONS

IN THE PRINCIPLES

OF

COOKING.

IN THREE PARTS.

BY

LADY BARKER,

*Author of "Stories About," "A Christmas Cake,'
&c. &c.*

London:
MACMILLAN AND CO.
1886.

CONTENTS.

PART I.

	PAGE
INTRODUCTORY	3

LESSON I.
THE CHEMICAL COMPOSITION OF OUR FOOD 10

LESSON II.
BREAD AND BEEF 18

LESSON III.
FISH 25

LESSON IV.
VEGETABLES 29

PART II.

LESSON V.
THE PREPARATIONS OF FLOUR USED AS FOOD 38

LESSON VI.

POTATOES AND OTHER VEGETABLES 44

LESSON VII.

MODES OF PREPARING BROTH OR SOUP FROM BEEF . . 51

LESSON VIII.

FUEL AND FIRE 58

PART III.

LESSON IX.

BOILING AND STEWING. 73

LESSON X.

BAKING, ROASTING, AND FRYING 79

LESSON XI.

BACON 86

LESSON XII.

THE GIST OF THE WHOLE MATTER 88

PART I.

THE CHEMICAL COMPOSITION, AND THE EFFECT UPON THE HUMAN BODY OF THE VARIOUS SUBSTANCES COMMONLY EMPLOYED AS FOOD.

FIRST LESSONS

IN THE

PRINCIPLES OF COOKING

PART I.

INTRODUCTORY.

The day has come in English social history when it is absolutely the bounden duty of every person at the head of a household—whether that household be large or small, rich or poor—to see that no waste is permitted in the preparation of food for the use of the family under his or her care. I am quite aware that such waste cannot be cured by theories, and that nothing except a practical acquaintance with the details of household management, supplemented by a conviction of the necessity of economy, can be expected to remedy the evil. At the same time, it is possible that ignorance of the fundamental principles of the <u>chemical composition</u> and of the <u>relative nutritive value</u> of the various sorts of food within our reach, added to the widespread ignorance of the most simple and wholesome modes of preparing such food, may be at the root of much of that waste.

Many excellent works have been written on household management and expenditure on both a large and a small scale, but I am not aware of any book so small as this, which exactly supplies the need I speak of, or which, laying other details aside, deals only with the subject of the preparation of food, and yet is not exactly a Cookery Book.

I shall attempt in this part to give in a condensed form the reasons why one sort of food is better than another, more nutritious, and therefore cheaper, and also why certain methods of preparing that food will cause it to be more easily digested, and render it more wholesome. It must be stated in this, the very beginning, that these "reasons why" are not the result of any crude theories of my own, but are drawn from a careful study of works upon the subject by practical chemists. Whenever the question is a vexed one, or learned doctors have agreed to differ upon it, I omit it altogether, confining myself entirely to the discussion of subjects upon which there is no doubt, and stating the results of years of patient study and incessant experiments as briefly and simply as I possibly can. Although it is perhaps somewhat alarming to come across scientific expressions in so unpretending a little book as this, still I must entreat my readers not to be scared away by words which are unfamiliar to them; and I may truthfully add my own experience to bear out the common assertion that the best and highest method of learning any subject will always prove the easiest in the long run.

Instead of helplessly wringing our hands and crying out about the high price of fuel and food, let us accept the present state of things as the inevitable and natural result of past years of extravagance and carelessness on our own part. The sooner we make up our minds that what we regretfully speak of as the "good old times" with their good old prices will never come again, the sooner we shall cease to look fondly back on a cheaper past, and brace ourselves up helpfully and bravely to face the increased cost of the necessaries of life. It is much more sensible to do this, instead of going on in our old ignorant way, buoying ourselves up with hopes of a shadowy millennium of butchers' meat, of a future day when carcases of Australian or South American sheep and oxen shall dangle in English shops. Believe me, that time is a long way off, and even when it does come there will be many more thousands of hungry mouths to be filled, so that the supply will only keep pace — even then rather lagging behind, as it does now — with the demand of the coming years. If fuel and food cost nearly twice as much at present as they did ten years ago, then surely it becomes our imperative duty to see how we can, each of us, according to our possibilities, make the material for warmth and cooking go twice as far as they have done hitherto. Nor in making such an attempt are we blindly groping in the dark, feeling our way step by step along the unaccustomed paths of scientific experiment. It has all been done for us whilst we were stupidly spending our capital, by men whose clear sight could discern the dark days ahead; men who have, many of them, gone to their rest, before the dawn of these dark days, but who have left behind them clear instructions how to make the most of certain necessary substances whose increasing value they foresaw twenty or thirty years ago. If, therefore, we have the common sense to avail ourselves of the results of these researches and experiments, which are still carried on day after day by worthy successors of the great practical chemists I speak of, it is quite possible we may so utilize their information as to make our available material go a great deal further. At present we all confess that the balance is uncomfortably adjusted, and a great many people are throwing a great many remedies into the uneven scales. Let us try a few grains of science, and a few more of common sense, and see what the practical result will be.

Before we proceed to do this, however, I should like to endeavour to disabuse my readers' minds of the idea that economy and stinginess are synonymous terms. In point of fact they are precisely opposite. An individual or a household habitually practising economy has a far wider margin for charity and hospitality than the shiftless people who never can keep a penny in their purses or a meal 'n their cupboards through sheer "waste-riff," as the north-country people call it. "Take care of the scraps, and the joints will take care of themselves," would be a very good motto in nine-tenths of our middle-class households, and the practical result of such a theory should be better food and more of it.

For my own part I have little hope of any real progress being made in the right direction until it shall have become once more the custom for ladies to do as their grandmothers did before them, and make it their business to acquaint themselves thoroughly with the principles and details of household management. In many cases there may be no actual pecuniary necessity for such supervision, but it would at all events serve the good purpose of setting an example, besides teaching servants the real good and beauty of a wise economy, a liberal thrift. So long as the world lasts, so long will there be a Mrs. Grundy; but if Mrs. Grundy can only be induced to go down into her kitchen and insist on a good use being made of sundry scraps and bones, and odds and ends which at present may be said to benefit no one, then will she deserve a statue in the market-place. If Mrs. A., whose husband's income may be one or two thousand a year, is able and capable to show a new cook how such and such things should be done so as to combine economy with palatableness, then will Mrs. B., whose income is barely a quarter of that sum, not consider it beneath her dignity to do so. If this movement is to do any good, it will have to be inaugurated by people whose social and pecuniary position makes them, to a certain extent, unaffected by the pressure which weighs so heavily on their poorer neighbours. And I am going to attempt, so to speak, to kill two birds with one stone; to persuade even rich people to insist on a due economy in the consumption of the necessaries of life, and to assure poor people that it is possible to make a good deal more of the scanty materials within their reach than they do at present. When I speak of inducing rich people to be economical, I have no culinary Utopia in my mind's eye, when millionaires will prefer to dine off cold mutton or to lunch on bone broth. What I mean is, that rich people can surely be made to understand that it is now-a-days absolutely a greater good to the commonwealth if their households are so managed that little or no material for human food can be wasted in them, than if they subscribed ever so liberally to all the great charities of London. It is just in proportion as people's minds are enlarged and their field of mental vision extended by culture and true refinement, that they will be able to perceive the importance of the question. For that reason I hope and expect that the warmest supporters of the attempt now being made by the National School of Cookery to teach the mass of the English people how to make the most of the material around them, will be found in the higher ranks of our society, and that from them it will spread downwards until it reaches the cottage where the labouring man is fed from year's end to year's end on monotonous and often unwholesome food, as much from lack of invention as from shallowness of purse.

Before ending this preliminary lesson I feel it incumbent on me to state most emphatically that I do not wish or intend to organize a crusade against cooks! In the course of nearly twenty years' experience of that class of servants, I can declare that I have found very little intentional dishonesty. Waste, extravagance, and bad management I have met with over and over again, but these evils have almost invariably arisen from want of opportunities of learning better, and I can scarcely remember an instance where there has not been an effort made to lay aside bad habits and acquire fresh ones. It is only too true, as dear Tom Hood says, that—

> "Evil is wrought by want of thought,
> As well as by want of heart."

So, if we can even teach our servants to think twice before they throw things into the pig-tub, it will be taking a step in the right direction.

If a cook and her mistress are at daggers drawn, each regarding the other as a foe to be distrusted, then, indeed, there is little real economy to be expected. But if a cook sees that her mistress is willing to give her fair wages for her services, and to consider her comforts in other ways, whilst at the same time the lady thoroughly understands *how* the cook's duties should be performed, the chances are that the servant will readily submit to be taught a thousand little helpful and comfortable ways. Such knowledge on the mistress's part is not incompatible with accomplishments and refinement of taste and manner, but it is not to be learned from reading this book or any other book. It can only come from study and a possibility of acquiring practical experience on the subject whilst the future matron is still a young girl; and if the scheme of the Committee of the National School of Cookery can be carried out according to their views and intentions, it will be a woman's own fault if in future her first visit to her kitchen be made as an inexperienced bride with a dozen years of apprenticeship before her ere she can venture even to make a suggestion to her cook, or dream of "tossing up" some little dainty dish with her own hands.

LESSON I.

THE CHEMICAL COMPOSITION OF OUR FOOD.

THE old German poet who wound up each verse of his famous drinking song by the assertion that "four elements intimately mixed, form all nature and build up the world," was not so far wrong after all. The jovial song-writer referred to his favourite formula for brewing punch; and according to him the world of conviviality was built up by lemon and sugar, rum and hot water.

Now, it is perfectly true that four elements go a great way towards building up the world; but, setting aside the question of brewing punch, they are called carbon, hydrogen, oxygen, and nitrogen. So universal is their presence in the living and growing parts of animals and plants, that they are always spoken of as "organic elements," and science has ascertained exactly the proportion in which each should exist in a healthy condition of the human body. That body is incessantly, but imperceptibly, undergoing a process which cannot be better described than by the expression of perennial moulting, only that, whereas certain animals cast off certain parts of their body—their skin, their hair, or their feathers—every year, we lose a portion of our weight every day; that is to say, we should lose it if we did not absorb through our lungs, the pores of our skin, and our stomachs, sufficient oxygen, carbon, hydrogen, and nitrogen, to supply the loss caused by the wear and tear of our daily life. There has even been an attempt made to prove that our vital organs are entirely renewed every forty days or so, but for this calculation there can be no really satisfactory data, although there certainly is constant loss and gain going on within us. The material for repairing this incessant waste which is the inevitable result of the activity of our nervous and muscular system, is not supplied alone by the starch, sugar, water, and fat, nor yet by the milk, meat, and vegetables we consume, but by a due combination of food material which shall ensure the proper proportions of albumen, fibrine, and caseine absolutely required by our changing frames. These are rather hard words, but their meaning will be quite plain if we take as familiar examples of the three indispensable ingredients, the white of an egg, a piece of lean meat, and a bit of cheese. Everyone can understand that, although these things contain the largest proportion of one particular substance, still there may be many other substances in which they are present, all together, and it is just to teach us this, and to explain to us why we should rather give our attention to procuring one form of food than another, that a knowledge of the elements of Practical Chemistry is useful.

In reading the accounts of the hardships and sufferings of explorers and travellers, we are often surprised to learn that first one member and then another of the expedition dropped down and died long before the supplies were actually exhausted. This is particularly noticeable in the account of Burke and Wills' attempt to explore the great plains of South Australia, where one by one the travellers died, not so much from sheer lack of some sort of food to eat, as from the unhappy circumstance of the only attainable food being utterly deficient in the ingredients without which the human body cannot be nourished. For instance, there was abundance of an alkaline plant on which the natives almost live at certain times of the year, and occasionally even a few fish were caught.

But these materials taken by themselves were so weak in life-supporting properties, that they failed to repair sufficiently the waste caused by severe exercise and exposure to the weather. A man may be starved to death, and yet scarcely feel hungry; that is to say, he may be able to put food into his mouth which will allay the cravings of his appetite, but which may not have the least power to nourish his body, so that he will die as surely as though he had nothing to eat.

Men's instincts are generally the surest guides, and however much we may have been disgusted to hear of such facts as of Esquimaux and Samoiedes living upon blubber and fat, and even eating 8 lbs. or 10 lbs. of flesh at a meal, Science teaches us that they were unconsciously adopting the very best means of keeping up the supply of carbon and oxygen, or internal warmth, which their cold climate rendered absolutely necessary. So in the same way we often see a sick person take a fancy to some curious kind of food, and perhaps begin to recover from the moment he was allowed to have it. The chances are that if we could bring all the practical chemists in the world into his sick-room, and they were to analyse the component parts of that particular food, and at the same time ascertain exactly which of the organic elements of human life was insufficiently represented in the patient's system, the result of their researches would go to prove that the sick man knew exactly what he wanted to build him up in health, better than anyone else.

Nature is our surest guide after all, only unfortunately our civilization has blunted our instincts, and rendered us more or less artificial, so that we can hardly tell what *is* Nature, and are obliged to call in the aid of Science to teach us. Those who live in hot countries do not require to provide their systems with internal warmth by means of food, and we shall generally find that they prefer a diet which will contain very little carbon. But it often happens that an Englishman travelling or living in such places will become terrified at his loss of relish for meat and heating food, and will fly either to his doctor for tonics, to his cook for pickles to incite his flagging appetite, or, still worse, to wine or brandy for stimulants to repair his imaginary weakness. Nature, thus thwarted in her arrangements, turns sulky, and the man falls ill, accusing the climate of the fault springing from his own ignorance and folly. In his own country he knows much better what is good for him; and in mixing bacon with his beans, or in taking, like the Irishman, cabbage with his potatoes, or, like the Italian, a strong kind of cheese with his maccaroni, he exhibits so many purely chemical ways of preparing mixtures nearly similar to each other in composition and nutritive value.

In the rudest diet, and in the luxuries of the most refined table, the main cravings of animal nature are never lost sight of. Besides the first taste in the mouth, there is an after-taste of the digestive organs, which requires to be satisfied if we want to arrange a perfect diet. It is not necessary that a food should yield every kind of material which the body requires to nourish it, for then one sort of food might be sufficient for the wants of man. Each sort must fulfil one or more of the body's requirements, so that by a wise combination the whole of its wants may be supplied. It is also to be borne in mind that our nourishment is not only the solid food which we actually take into our stomachs, according to the popular idea on the subject, but comprises the water we drink and the air we breathe. But as these pages should treat simply of the nourishment for our bodies, which nourishment must needs be submitted to the action of fire, it is only with the cooking of food we have to deal.

In considering the question of the best and cheapest food, and the most wholesome mode of cooking it, we must keep steadily before us the principle, that it is not the quantity of food received into the human body which nourishes it, but the proportion which can be digested of such food. All else is sheer waste—an encumbrance worse than useless—whose presence clogs and throws out of gear the delicate mechanism appointed to deal with it.

It is generally agreed by scientific chemists, that in casting around for something like a form of food which could be taken as a type of all others, there is none so perfect as milk. During the period when the young of animals as well as of human beings are fed entirely on milk, they grow very rapidly in the size of every part of their bodies. From this we infer that milk must contain *all* the essentials which go to build up muscle, nerve, bone, and every other tissue. The first lesson we learn from taking milk as an example of perfect natural food, is that there should be a certain proportion of liquid mixed with the substances we consume as food, though, as the animal attains its full size and there is only waste to be made up, not growth to be provided for, the necessity for the liquid form of food diminishes.

Of the flesh-forming substances contained in milk, caseine is the most important, and in the largest proportions; therefore it is with milk in the form of cheese that it can best be dealt with as human food in this place. Now, there is a popular theory that cheese is unwholesome, and it certainly is an indigestible substance, but still it need only be avoided by those who suffer from weak digestions. The hardworking man who labours with his muscles in the open air, and whose stomach is in the best possible condition to digest his food, does wisely to spend, as he generally does, what little money he may possess in cheese, for cheese contains nearly twice the quantity of nutritive matter he would get in the same weight of cooked meat. Even with delicate feeders, a small quantity of cheese taken with other food facilitates digestion, for caseine is easily decomposed or put in a condition which causes other things to change. When, therefore, we eat a piece of cheese after a meal, it acts like yeast in bread, and starts a change in the food; for the chances are that the stomach in trying to digest the cheese will digest the rest of its contents at the same time. The mouldy cheese which some people's instinct leads them to prefer, acts more quickly in this way than fresh cheese. When cheese is spoken of as a nourishing article of food, especially to those who labour in the open air, it is only cheese in which the cream has not been previously separated from the milk, for the actual nutritive value will depend on the amount of butter material left in it. The cheap skim-milk cheeses of South Wales yield so little nourishment in this respect, that they are of but slight value as flesh-formers, whereas the rich cheeses from Cheddar, Stilton, and Ayrshire are not only infinitely cheaper than meat, but are also very nourishing.

It will perhaps only be necessary to take bread and beef as samples of food which contain in themselves every element required to build up the human frame, to repair the daily waste, and to preserve all the conditions of perfect health. The generality of mankind have found out the value of these substances for themselves without the aid of science; but it may be as well to learn something about bread and beef, for the simple reason that as we cannot always, under all circumstances, make sure of having them as food, we may be able to select those substances which come nearest to them in nutritive value, if we understand the component parts which make them so important.

LESSON II.

BREAD AND BEEF.

NATURE is always busy cooking inside us. She is ever separating, arranging, and making the best of the heterogeneous substances we give her to deal with, and it is as well to find out what materials are the easiest for her to manage, and so learn to economize her forces to the utmost. Of all the food used to repair the incessant waste caused by muscular exertion in the open air, bread and beef, as we have already remarked, best fulfil the needs of the human system under those conditions; and we will first look at the chemical composition of bread.

It is needless to trace the growth of wheat before it arrives at the mill to be converted into flour, but when it reaches that stage it comes within the limits of the inquiry which we propose to ourselves. Wheat is practically divided into two parts: the bran or outer covering, and the central grain or fecula; and the object of the miller in the preparation of flour is to mix the qualities as above mentioned so as to suit his market, and either to separate the bran entirely or partially from the grain, or to leave the whole in flour. According to the quality of the grain and the amount of the husk left in it, the value of the flour varies, and it is divided into four classes: the "fine households" or best, "households" or "seconds," brown meal, and biscuit flour; and the value must chiefly depend on the estimate which is formed of the nutritive proportions of the different parts of the bran.

Many people say, vaguely, "Oh, brown bread is more wholesome than white"; but it is impossible it can be more nutritious, though it may be more palatable; for the outer part of the bran is glazed over with a layer of flint which is quite indigestible. At the same time it must be acknowledged that our practical experience teaches us that, although the stomach may find it impossible to assimilate bran itself, yet the presence of bran in bread stimulates the juices of the stomach to greater activity, and therefore, like cheese, promotes the digestion of other things. To a delicate organization it would probably act as an irritant, and therefore its use should not be persisted in unless there is absolutely no disarrangement of the digestive system. However finely the *outer* bran may be ground, it still remains innutritious, but the *inner* husk possesses great value from the large proportion of nitrogenous matter which it contains. The whiteness of the flour is not always a test of its purity or nourishing powers, as in cases where the flour from red wheat has been most thoroughly sifted or "bolted," it will still keep a darker tinge than even "seconds" flour obtained from white wheat, though the red wheat remains the most nutritious.

It is an instance of what I have before remarked about the instinct which guides our choice of food, that the navvies, who work perhaps harder than any other men in the world, make it a point to procure the very best and purest and most expensive wheaten bread. It is always the first thing thought of in settling to a job of work in a new place, that these men should be able to get the finest wheaten bread to eat. In making this proviso they are really guided by principles of true economy, for in their case the necessary waste of tissue is so great that they cannot afford to take into their stomachs any superfluous matter which will not nourish their bodies. And we will presently see *why* pure wheaten bread is the most nourishing of all the cereals, although there are other forms in which wheaten flour might be used with advantage, such as when made into maccaroni or sifted into semolina.

In other countries, where wheaten bread is not the staple article of food, it is curious to notice how those who have to work hard in the open air have struck out substitutes for themselves which contain ingredients as near to wheaten bread in chemical value as can be procured. Thus the miners of Chili, whose lives are very laborious, feed on beans and roasted grain; whilst some Hindoo navvies found their physical powers too low to do a good day's work when engaged in boring a tunnel, until they left off eating rice and took to wheaten bread and flesh. But the wheat grown in a tropical country is never of much value for nutritive purposes, nor yet that grown in a cold one. A hot summer in a sunny clime lying within the temperate zone produces the best grain—that is, grain with the least proportion of water and the greatest of nitrogen. Rice flour possesses so much less nitrogen than does wheaten flour that its nutritive value is a good deal lessened, and in countries where it is the staple food, a very great deal has to be produced and consumed to afford the inhabitants anything like a sufficiency of nourishment. The innutritive quality of rice is naturally the reason why a scarcity of that food causes such fatal results in an apparently short time. The people who habitually eat it have already brought their vital powers to so low an ebb, that a very small diminution of nourishment suffices to lower the life-supporting standard beneath the possibility of existence. The chief reason why wheat, and indeed all the cereals, are of such primary importance as food, is, that whilst nitrogen is absolutely indispensable to the animal body, it cannot be produced out of substances which do not contain it. The same is true of carbon, but we must look to flesh to produce that. The chief ingredients of our blood contain nearly 17 per cent. of nitrogen, according to Liebig, and he was also convinced that no part of an organ contains less than the same proportion of that clementary body. The nitrogenous principle in wheat is called gluten; but it is the *cerealin* which acts as a ferment and assists in the digestion of the other substances.

In wheat this is what we find—water, gluten, albumen, starch, sugar, gum, fat, woody fibre, and mineral matter, all in certain proportions, but there is a great deal more starch than anything else. Next to starch comes gluten, and we must remember it is in that ingredient the nitrogenous principle lurks. If these component parts are again classed, the result will be that wheat stands first as a "force-producer," and second as a "flesh-producer;" so, as strength is of more importance to the navvies than flesh, they may well be excused for being so particular about their bread. In another place we will speak of the simplest and best modes of making wheaten flour into bread. Now we must pass on to beef, and try to show why our national love of this particular form of flesh-food has had its origin in an instinct of what was best to keep ourselves in good working or fighting condition.

Although bread actually produces fibrine, still it is best if we need only look to it for gluten, albumen, and so forth, and depend upon flesh for fibrine, where we shall find it ready-made to our hand (or, should I say to our mouth?) in the fibres of the meat. Of all the forms of meat used for human food, the flesh of the ox is that generally preferred where there is any choice in the matter, and it is certainly both nourishing and easily digested. In comparing the nutritive value of different kinds of meat, we must distinguish between fat and lean, and the amount of nourishment is in proportion to the fat or lean of the meat. Fat (that is, carbon) generates heat, but lean generates heat and forms flesh as well, for in lean flesh all four "organic elements" are well represented. In both mutton and pork we get so much fat that the actual nourishment contained in the same amount of beef (unless exceptionally fattened) is greater, and it is also the fullest of the red blood juices. Besides this, the loss in cooking beef is much less than in cooking mutton, owing to the greater solidity of the flesh and the smaller proportion of fat. "It is quite certain," says Liebig, "that a nation of animal feeders is always a nation of hunters, for the use of a rich nitrogenous diet demands an expenditure of power and a large amount of physical exertion, as is seen in the restless disposition of all the carnivora of our menageries." Hence it follows that for those whose daily toil necessitates an expenditure of power, it would be the truest economy if they were to endeavour to supply the waste of their muscular system by ever so small a quantity of true flesh-forming food, instead of being contented with a larger meal of a less nourishing description, washed down by beer or spirit, which contains no real nutritive worth. Malt and alcohol possess narcotic and stimulating properties, and do no harm in moderation—indeed, to the weak or aged they are of incalculable value. But a strong, healthy labouring man would keep himself in much better working order if he economized his beer and increased his animal food.

I have seen with my own eyes a very forcible illustration of this truth in the working man of New Zealand as he existed some years ago. In those days beer and spirit used to be almost unknown except in the young colonial towns, and the early settlers up the country lived entirely on bread and mutton, for even potatoes were a rare and precious delicacy for the first half-dozen years. Such a splendid physical condition of the human frame it had never before been my good fortune to behold. Everyone looked in the perfection of health: clear complexions, bright eyes, and active limbs which seemed not to know fatigue, were the result of many years of a compulsory and much-abused diet of bread, tea, and mutton. When I say tea, it was really only used as a stimulant or for warmth, for cold water was the universal beverage. People might grumble, but they throve, and the generation whom I saw growing on that diet from childhood towards man's estate might challenge the world over to produce their equals for vigour and strength.

Perhaps it is rather "bull"-ish of me to insist in one page upon beef, like motley, being "your only wear," and then in the next going near to show that mutton does just as well; but, seriously, one has only to turn to Sir Francis Head's account of his ride across the Pampas, to learn how much exertion can be supported upon dried lean beef. It is not only, as Sir Francis says, that he endured enormous and incessant fatigue solely on this beef diet, but that months of such fatigue left him in splendid physical condition, able to do anything or go anywhere. To reconcile the two theories, however, I must add that the gallant veteran confesses his beef diet rendered him somewhat lean and ill-favoured, and that he did not look so handsome and well as my mutton-fed New Zealand colonists used to do.

LESSON III.

FISH.

In many parts of the coast of our sea-surrounded home, fish is, from necessity, the staple food of the inhabitants; and although whole districts in other parts of the world, such as Dacca, the Mediterranean coast of Spain, &c., are fed almost entirely on fish, our business lies only with our own people. There is no doubt that fish, even the red-blooded salmon, should not be the sole nitrogenous animal food of any nation; and even if milk and eggs be added, the vigour of such people will not equal that of a flesh-eating community. But of all kinds of animal food, the fresh herring offers the largest amount of nutriment for the smallest amount of money, and this statement is the more curious when we think of the turtle, which is produced in such enormous quantities on the shores of the West Indian islands, as well as the estuaries of the Indian coast. Although the flesh of the turtle is palatable and wholesome, it possesses a cloying peculiarity, insomuch that, after a year or two, Europeans will suffer hunger to the verge of starvation rather than touch it. Perhaps this repugnance may be an instinct arising from the fact that the phosphoric fat of the turtle renders it difficult of solution in the digestive juices, and therefore its really nutritious properties are counteracted by this superabundant richness.

So we see that the balance has to be very nicely adjusted: the old proverb, "If a little of a thing is good, a great deal is better," does not hold good at all with our food. We have to take great care that, according to the means within our reach, that supply of the proper proportions of the organic elements which are as necessary to our bodies as fuel to a fire, should be kept up. In fact, food is to our body exactly what fuel is to a fire. If we choke up the range or stove with dust and bricks, the fire will go out; and so, if we persist in supplying the furnace of our life with materials which it cannot possibly assimilate, or use as fuel, the fire of our lives will die out. If people understood, or would even try to understand —and it is not so difficult as many things uneducated people learn quite easily—why certain kinds of food produce certain conditions of the human frame, there would be far less disease.

The great mistake is to think that actual want of money is at the root of the bad food of English labourers. It is not so at all. I do not deny the poverty nor the toil requisite, alas! to obtain even the scantiest meal; but anyone with any practical experience of the very poor of our own country will agree in the assertion that perhaps half of that pressure is removable by education in the art of making the most of things. I have often seen a poor woman who had been complaining to me of the scarcity of fuel, or the want of food, prepare to light her fire, cook her husband's dinner, or bake her bread, in the most recklessly extravagant manner. So with fish. How often at the time of the Irish famine were the charitable English public startled by hearing that people were starving on a coast swarming with fish? If it had been possible to teach the poor ignorant sufferers, that although there was not quite so much nourishment in fish as in meat, still it would have made a palatable and wholesome addition to their starvation diet of Indian maize, much distress would have been warded off.

The flesh of fish contains fibrine, albumen, and gelatine in small proportions, and fat, water, and mineral matter go to make up the rest of the component parts. It is curious to find the difference of fat in some fishes, especially mackerel, which possesses a very large proportion, herrings coming next (some people say first), but at all events they both should be cooked in such a way as to get rid of as much of this fat as possible. Enough will remain to make the fish nourishing, but if there be too much fat it renders fish indigestible. This danger needs to be particularly guarded against with eels. Haddocks, whiting, smelts, cod, soles, and turbot are all less fatty, and consequently more digestible, than such fish as salmon, pilchards, sprats, and mackerel. Raw oysters are more digestible than cooked ones, because the heat coagulates and hardens the albumen at once, besides making the fibrine too solid, and rendering it less easy for the gastric juices to dissolve.

We must bear in mind that the flesh of all fish *out of season* is unwholesome, and often makes people ill. I am afraid Mr. Frank Buckland and other true lovers of pisciculture would view the sufferings of such depraved *gourmets* with great indifference, and it is, indeed, most shocking to the food-economist to read of the shoals of baby soles an inch or two long, of diminutive oysters, of the ova of the cod, the roe of the salmon, and of the fry of the herring, which are brought to our markets and readily sold in spite of vigilant bye-laws.

It is not possible in this place to deal with the subject of cooking fish : cooking it in such a manner that the fat which renders it often unwholesome shall be eliminated, and the nourishing and gelatinous portions of the fleshy substance made the most of.

LESSON IV.

VEGETABLES.

I FEEL that I cannot begin this chapter better than by quoting what Dr. Letheby says on the subject:

"Primarily, *all* our foods are derived from the vegetable kingdom, for no animal has the physiological power of associating mineral elements and forming them into food. Within our own bodies there is no faculty for such conversion; our province is to pull down what the vegetable has built up, and to let loose the affinities which the plant has brought into bondage, and thus to restore to inanimate nature the matter and force which the growing plant had taken from it."

It is thus plain that the beef and mutton we eat derive their fibrine, gluten, and all other necessary ingredients from the vegetables on which the oxen and sheep have fed, though such food does not apparently contain any of these substances. It is a curious suggestion which I have often met with, that if a vegetarian family lived in accordance with the rules of one of their own peculiar cookery books, each member would actually consume half an ounce more animal food a day than a man would do who lived according to the usual scale of diet.

Vegetables are aliments which dilute the blood, and

contain more salts than albumen. They convey very little nutriment to the blood, as we may see in the feeble muscles of tropic-dwellers who feed almost entirely on vegetables. On the other hand, they are of great service, first in the digestive canal, where they dissolve the albuminous substances of the meat, and afterwards in the blood itself, where, if they do not actually nourish, they yet keep the albumen and fibrine in a liquid state, and enable those substances to perform their proper functions more vigorously. Of course the cereals would naturally stand first in a chapter on vegetables, as they, of all the products of the vegetable kingdom, are the most depended upon by man for food. As, however, wheat, which is the principal cereal of England, has been noticed in another chapter, we may as well proceed to examine the nutritive properties of other vegetables. In such an inquiry the potato comes first, for, owing to its large proportion of starch, it is the most actually nourishing of all vegetables. This starch is transformed into fat by the digestive process, and if potatoes could be eaten with a sufficiency of white of egg, their nutritive value would be brought very near the meat standard. Other roots and tubers contain a larger proportion of sugar, and there is even fat present in some of them, but none are so rich in this nourishing starch as the potato. A man may, and probably will, look fat and rosy on a potato diet, yet his muscle will not be in first-rate condition, nor will he be able to endure prolonged fatigue. In spite, therefore, of the comparative low price of potatoes, they are not the most economical food for a labourer, nor can he depend on their nourishing starch alone to provide him with the requisite bodily strength. All succulent vegetables are anti-scorbutic, and since the potato was brought into use as a daily ration in the fleet (not a hundred years ago), scurvy has gradually died out. If there is any difficulty in providing potatoes—for during long voyages, when crossing the tropics, the potatoes will begin to grow, and so become unfit for food—lime-juice is the next best substitute, for it contains most of the chemical ingredients which go to make the salts of potash found in all fresh vegetables, but which is specially present in the potato. It has often been pointed out that there is really no excuse for scurvy now-a-days, for potatoes, cabbages, turnips, and carrots can be pressed into a very small space, and yet carry their potash about with them. Indeed, this process has lately been carried to great perfection. Other vegetables are less actually nutritious than the potato, and the palate grows sooner tired of them, but yet one hundred pounds of potatoes contain barely as much nitrogenous matter,—that is to say, positive nourishment,—as thirteen pounds of wheat.

As the wholesomeness and digestibility of vegetables depend much on how they are cooked, it is perhaps useless to enter here into a longer explanation why vegetables, though they constitute the entire food of animals whose flesh contains the highest forms of nourishment, will not, of themselves, supply man with the food he requires to keep his muscles strong and vigorous. In the countries where the inhabitants are compelled by the necessities of the climate to live chiefly on them, Nature is so bountiful that she does not call upon man to cultivate the ground as we are obliged to do. Therefore, it stands to reason that in a climate where severe manual labour is necessary to produce food, a diet of a muscle-relaxing, fat-forming nature is a very poor economy.

PART II.

THE BEST MODES OF PREPARING SOME SORTS OF FOOD FOR USE, WITH A SIMPLE EXPLA-NATION OF THEIR RESPECTIVE ACTIONS.

PART II.

REMARKS.

THE very first principle of cooking is cleanliness. No skill or flavouring can make up for the lack of it, and if it be present, there is good hope of every other culinary virtue. But cleanliness is an elastic term, and I wish it to be clearly understood that I would fain stretch its interpretation to the utmost limit. Even the sacred frying-pan would I ruthlessly scour, all unheeding the old-fashioned, and, let us add, dirty axiom, that it should be left with the fat in it. It is quite true that the fat which has been used to fry potatoes, or fritters, or anything *except* fish, may be poured out of the saucepan into a daintily clean basin or empty jam-pot and used again and again, but I would have every cook taught to clean her frying-pan thoroughly every time she uses it. The fat in which fish has been fried should *never* be used for frying anything else, and an economical housewife will take care that the fish is fried last. I have sometimes been met with the assertion that it is too much trouble and takes too much time to keep everything in a kitchen as clean as it ought to be kept. To that I reply, that if a girl be brought up by a tidy mother or mistress to understand and appreciate the value and beauty of cleanliness, she will never be able to endure any other state of things. I declare that I have observed greater dirt among the saucepans and a deeper shade of black over everything in kitchens where neither poverty nor want of time could be pleaded in excuse, than in a place where one pair of willing hands has had to keep the living-room of half a dozen people tidy.

I am not sure that I do not detest surface-cleanliness, with its deceptive whiteness, more than genuine honest dirt about which there is no concealment, for the sham snowiness is apt to throw youthful housekeepers off their guard. For their encouragement I can assure them that it is not such a superhuman task as it appears to see that everything under their sceptre is kept scrupulously clean, for the advantages of cleanliness over dirt are as patent as light over darkness, and ninety-nine servants out of a hundred will soon come to acknowledge this themselves. People of all ranks and classes differ in this respect according to their instincts and training, and in many a fine house a dirty cook would find things more after her own heart than in a two-roomed cottage.

Let us, for a moment, take the case of a girl who has been a housemaid or nursemaid in a small family, and who marries a decent young artisan earning from 15*s*. to 25*s*. a week. Here is enough money for comfort *if* the wife knows how to manage and is clean and tidy in herself. How far will that, or twice that sum, go if she be an ignorant slattern? The chances are that such a girl knows absolutely nothing of cooking, and that she will have to arrive at even the smallest amount of such knowledge through a long series of unpalatable meals and wasted food. Perhaps it may be years before she attains to the production of any dish which can fairly be called wholesome or nourishing; but surely she is not to be blamed for her ignorance. She has gone straight from her school to a situation whose duties have never taken her into the kitchen, and she finds herself at twenty-five years of age at the head of a working man's home, with no more notion of how to manage their income comfortably than if she were an infant. She has hitherto had no opportunity of learning how to cook; but if she has been taught to be thoroughly clean and tidy in her habits and ways, she may rest assured that half the battle is won. The other half, the National School of Cookery at South Kensington steps in to help her to win, and it is to be hoped that in due time, by the establishment of branch institutions all over the kingdom, by means of lectures and demonstrations (for cooking cannot be taught by theory), any young woman in such a position will know where to go if she wants to learn how to cook the food her husband's wages enable her to provide. But *cleanliness* she must teach herself, and practise it diligently in her little kitchen, for without it she can never be a good cook, no matter how successful she be in the matter of bread, or how deftly she may handle her frying or sauce pan.

LESSON V.

THE PREPARATIONS OF FLOUR USED AS FOOD.

It is well known that so far as actual nutritive power goes, both oats and barley, to say nothing of maize, rye, the millets, and rice, contain as much (oats, indeed, more) valuable material for the maintenance of the human body as wheat does; that is to say, they all contain certain proportions of starch, protein, or the nutritive ingredient, represented by oily or fatty matter, besides sundry saline particles. All these are indispensable to the building up of the human body. Why then do we find wheat more cultivated and used in greater quantities by all the civilized nations than any of the other cereals? The only reason can be that wheaten flour alone, of all these farinaceous foods, will make fermented bread.

I used at one time to think that bread-making must be the very simplest thing in the world, but when I came to be face to face with flour and yeast I found it was not so easy a matter to produce light good bread. These pages are not written therefore for the instruction of bakers or those fortunate people who have learned, at an age and under circumstances when learning is easy, how to make bread, but with the hope that they may prove ever so slight a practical help to those who are as profoundly ignorant as I was, not so long ago.

First of all the yeast has to be thought of. When near a town this thorn in the path of the anxious bread-maker is removed by the facility with which brewer's or ready-prepared baker's yeast can be procured. Brewer's yeast is simply the scum which rises to the top of the malt during the process of fermentation, and is of no use to the beer, or wort. The brewer is therefore glad to dispose of it, and the baker takes it off his hands. But he does not put it raw into his bread. A special ferment is first obtained from mealy potatoes, by boiling them in water, mashing them, and allowing them to cool to a temperature of about 80° of Fahrenheit. Yeast is then added to them, and in a few hours they will get into a state of active fermentation with a sort of cauliflower head. Water should now be gently poured into this mixture, and it must be strained, after which a very little flour should be lightly sprinkled into it. In five or six hours the whole will rise to a fine *sponge*, when more water must be added, and a little salt, and then the yeast is fit to use. It may now be bottled, but it is not advisable to make a great deal at a time. On account of the fermentation, yeast-bottles can only be kept from bursting by plugging their mouths with soft paper or cotton-wool. If neither the fresh yeast from the brewers (which will not keep by itself for more than a day or two) or the dried yeast, which keeps a long time, can be obtained, then it will be necessary to boil some dried hops in a very little water, put some sugar to them, and add this compound when in a state of fermentation to the mashed potatoes instead of the brewer's yeast.

Having procured or made the yeast, the next thing is to put the flour in a large tin milk-pan, make a hole in the centre of the soft white heap, and pour in a small cupful of yeast mixed with a large cupful of warm water. A little of the flour is stirred in to this liquid so as to make it rather more of a paste, and then the whole is covered with a clean cloth and set to *work* during the whole night. Great care must be taken not to put it in too hot a place, as it will become dry and crusty in the morning, and make heavy, tasteless bread. On the other hand, if the temperature be too low, the flour will be dull and cold, the mixture will not have penetrated it, and the bread will not rise. But, supposing that the happy medium has been hit, and that the gas contained in the yeast has made its subtle way among the flour, then more water must be added by degrees and a very little salt. The whole mass should then be lightly kneaded by *very* clean hands, and when it has attained a certain elastic consistency it should be quickly cut into separate portions, dropped into well-floured tins (only half fill them with the dough), which must instantly be placed in the oven. The oven should be fairly hot to begin with, and its heat increased until the end. From time to time a clean knife should be thrust into the loaf; if it comes out with a tarnish on the bright blade, as though it had been breathed upon, then the bread is not sufficiently baked, and there is no use in taking it out of the oven until the knife can be readily drawn out with a perfectly undimmed surface. The real art of bread-making consists in the dough not being too stiff at first to resist the entrance of the gas, nor too soft to permit the gas to pass through it quickly. It should also be sufficiently kneaded so that the gas may become well distributed throughout the mass, yet not over-kneaded, in which case a good deal of it will have escaped, and the bread will consequently be heavy.

The difference between biscuits and bread is that there is no yeast in the composition of the former; they are also for the most part unleavened and very highly dried. Though valuable as a temporary substitute for bread, they can never be so wholesome from the absence of the water which is absorbed in the process of drying or baking. Biscuits should invariably be taken with ever so small a quantity of liquid, for by themselves they either absorb too much fluid from the juices of the stomach, and so produce indigestion, or they fail to obtain as much fluid as they require from those sources, and therefore remain a long time undigested. Cakes are made by the substitution of soda or carbonic acid for yeast, and the addition of sugar, fat, and eggs. Of all these materials the sugar is the wholesomest and should be the most freely used. The other ingredients are more difficult of digestion.

Before leaving the subject of bread, it will be as well to notice the extraordinary difference between batches of bread. It is no reason because a household receives excellent bread one week—either from the baker's shop or its own kitchen—that the next week's baking will not be heavy and bad. This is because we trust so entirely to the good old rule of thumb in our kitchens, scorning to make the temperature of the oven a certainty by means of a thermometer. Half, and more than half, of the hard baking and the over or under boiling and frying with which we are afflicted arises from the extraordinary prejudice which exists against the daily use of this indispensable little instrument. It is the only reliable way of making sure of the oven, or the water, or the fat being of exactly the right temperature; and yet what cook who "respects herself" would at present deign to use a thermometer, still less even a charming little contrivance which has been invented specially for her use, and is called a frimometer?

But to touch upon some of the other uses of flour. We are apt to look upon macaroni as a luxury for the tables of the rich, when it is really so low in price that it is within the reach of those who have any choice at all as to what they shall eat. It is considered a foreign composition, unworthy to take a place among the more solid flesh-formers dear to the heart of the Englishman; but if he understood what it is made from, he might perhaps modify his contempt for one of the most nourishing and wholesome forms in which he can eat wheaten flour. Macaroni, then, is made by the simplest imaginable process, and there is no reason in the world why its manufacture should not be carried on in· England, as indeed it is. The finest wheaten flour is made into a peculiar smooth paste or dough, and afterwards driven through a cylinder which cuts it into ribands or tubes. Wheaten flour contains, of course, precisely the same amount of nourishment, whether it be made into bread or into the *pasta* from which macaroni is cut; but whereas bread can scarcely be cooked again (except as toast), there are many ways in which macaroni can be dressed so as to form a delicious food. Simply boiled with milk and a little sugar it would be a wholesome and agreeable change in children's diet, and we must remember that for children who are born with soft bones—that is, with too little phosphate of lime in their bones—a diet of wheat will tend, more than anything else, to form this deposit. When I say wheat, I include macaroni therefore, and semolina, which is the very small grain left after grinding wheat in a coarse mill. Such a mode of grinding gives but a small proportion of flour, and a certain larger residue of coarse flour or fine grains, and these grains are known as "semolina." They are chiefly obtained from the most nourishing of all the wheats, the red-grained wheat grown in Southern Europe, and especially in the Danubian Principalities.

LESSON VI.

POTATOES AND OTHER VEGETABLES.

ALTHOUGH it is rather a departure from the plan I pursued in the First Part to speak in this lesson about potatoes, it is natural to me to do it, because, so far as my practical experience—which was once *in*-experience, remember—goes, it is almost as difficult to boil a potato properly as to bake good bread. In the first place, we have one of the highest chemical authorities on our side for saying that on both wholesome and economical grounds potatoes should always be boiled *in* their skins. They do not look quite so well if they have to be peeled afterwards, but not only is the actual material wasted by the process of peeling—especially where there are no pigs to eat the peelings—but a great deal of the starchy substance, which is exactly what makes the potato so nourishing, is wasted. In roasted or baked potatoes, which have been peeled before cooking, the loss in weight from the skin and the drying is actually a quarter of the whole. It is curious to learn that potatoes which come to us from the bog lands of Ireland are far less watery and produce more starch than those which are grown on the dry, light soils of Yorkshire. This innate dryness is one reason why the Irish potato contains so much more nourishment than an English one. The potato was first grown by Sir Walter Raleigh in his garden at Youghal, in Ireland, and it is not much more than a century since its cultivation became general in England. The first potatoes grown in England came from a ship wrecked on Formby Point, near Liverpool. The tubers were planted by chance on the soil close by, which closely resembled that of Ireland, and no part of their new home has ever suited them better. The potato, though, as we have seen, of a certain appreciable value as a flesh-former, is not to be depended upon entirely as a force-producer, for the proportion of water in 100 parts is 75·2. Next to water, its peculiarly nourishing starch is most largely represented, and stands at 15·5. From this starch also a *pasta* can be made which gives a fair macaroni, but of course the advantages of the wheaten paste would be absent.

In ordinary kitchens where a steamer is used, the process of boiling a potato is easy enough, and that dry mealiness dear to the heart of a good cook can be reckoned upon. But if only a saucepan be attainable, then, having well washed—nay, even scrubbed and *brushed*—your potatoes, put them into it with *cold* water; add a little salt when the water boils; at first it should only be allowed to boil slowly, but it may boil as fast as you like during the last five minutes. Some varieties of the potato can be cooked much sooner than others; there is often the difference between them of twenty minutes and three-quarters of an hour. From time to time they must be tried with a fork, which should go in freely when they are sufficiently boiled. The potatoes being now cooked enough, pour off as much water as can possibly be got rid of. Sprinkle a little more salt, take off the lid of the saucepan and set it on again in such a manner that the steam can escape, but keep the saucepan for a few minutes on the oven to dry the potatoes thoroughly. The saucepan should be lightly shaken from time to time to prevent the potatoes sticking to the bottom. Then serve either in a wooden bowl, with a clean cloth or a napkin, or else in a dish with perforated holes in the cover so that the vapour can escape. If potatoes form the principal diet of a family, eggs should be added where practicable, and milk, or dripping, or any sort of fat, as the potato itself is very deficient in albumen and fat.

Next to the potato, the cabbage is the most widely cultivated of all vegetables, yet it is far inferior to the others in the nutriment contained in a given weight. In point of value the parsnip ranks next to the potato as a flesh-former, and possesses six per cent. of carbon. Parsnips are followed closely by carrots and onions, though the latter are principally used as a relish. But all vegetables are chiefly valuable for their anti-scorbutic properties, and as a flavouring for insipid food. Lentils are particularly nutritious, and the food sold under the name of "Revalenta Arabica" is only the meal of the lentil after being freed from its indigestible outer skin. In peas we find a great deal of caseine; hence, in an analytical table they rank next to wheat as a flesh and force-producer, whereas we should find the other vegetables relegated under the head of "Non-nitrogenous substances," that is to say, substances which, taken by themselves without milk, butter, or fat of any kind, are absolutely incapable of producing either flesh or force. In Ireland it is the milk taken with the potato which makes it so nourishing. If potatoes were eaten quite alone, the consumer would need to eat an enormous quantity to keep himself in any sort of condition, and he would never be able to do any amount of real hard work in the open air.

It is quite certain that sufficient value is not attached in England to the importance of the cultivation of vegetables. If a few leeks or sweet herbs, a row of potatoes, or a dozen cabbages, were planted in many a tiny spot beside a cottage door, which spot at present is but a puddle or a down-trodden mass or caked mud, the hungry mouths inside would stand a better chance of being filled. When a poor woman has to go with her pence in her hand and buy every onion or potato or sprig of thyme which she wants to improve the flavour of the family meal, the chances are she will look upon them—and very justly, too—as luxurious additions to the bill of fare, and do without them as much as possible. All over France the poorest peasant has her "flavourings" close to her hand; and it is difficult to over-estimate the boon which a few common vegetables and herbs are, when used to assist in converting a scrap of bacon, a bone, and a little pea-meal into a warm, comforting, nourishing mid-day meal.

Mr. Ruskin attaches great importance to the cultivation of the land—the making the best of every inch of our own native soil; but I fear he wants to try experiments, and grow all sorts of curious things in every conceivable part of the British Isles, whereas I only confine my ambition to those little shabby nooks and odds and ends of ground which lurk around stray cottages, whose occupants evidently prefer sitting in the tap-room of the "Chequers" to digging for an hour in a scrap of garden morning and evening. Perhaps, if, in time, we are able to show the working man how enormously his culinary comfort can be increased by a little vegetable flavouring, he may take to planting and cultivating even a square rood of ground, if that be all he can call his own. I say nothing of the gain to health, for that is so easily ascertained by his own or his neighbour's experience. The seeds of common vegetables are very easily procured—in fact, they can almost be had for the asking; and, at all events, one day's beer-money would go a long way towards keeping a family in onions for a year if laid out in seed. A little soup or stew thus flavoured without extra expense, would surely be a vast gain on the hunch of dry bread and mug of weak, cold coffee, which I have often seen a labourer eating for his dinner. Then there only remains the trouble to be considered; and a lazy man will have to make twice as much exertion in the long run to keep body and soul together.

I repeat: it is not actual money which is absolutely wanting in such cases. It is that the few pence are generally laid out in the most improvident way— in a way which becomes gross extravagance when it is contrasted with what the same pittance would produce if properly managed. I have no hope of this little book, or any other book, great or small, working a miraculous and thorough reform, and converting every cottage in the country into a smiling abode of peace and plenty. What I *do* aim at and look forward to is, first, to arouse attention to the subject in those whose social rank is *above* that of the hand-to-mouth working man; and next, to induce rich people to take as much trouble and spend as much money in providing their servants and workmen with the opportunity of learning *how* to cook their food, as they now do in teaching them and their children to read and write.

Mr. Ruskin, in his "Fors Clavigera," insists very strongly that in his model farm, his land bought out of the proceeds of the "St. George's Fund," every girl shall be taught "at a proper age to cook all ordinary food exquisitely." But I would go a step beyond, and I would have every boy taught also. I don't know about the cooking exquisitely! I should be satisfied, at first, if every boy and girl could be taught to cook even a little. For a knowledge of cooking, at all events in its simplest form, appears to me to be every whit as necessary for a man, if he is to move about the world at all, as it is for a girl. If the man does *not* move about, and is fortunate enough to marry a girl trained and taught cooking either at Mr. Ruskin's model farm or at the National School of Cookery, then he may forget, or lay aside, his culinary lore as quickly as he pleases! But if he emigrates, or enlists as a soldier, or does any of the hundred and one things which men are obliged to do in these busy days, the chances are that he will find ever so slight a knowledge of cooking a very great boon and blessing to him.

One thing is very puzzling to me, though I know not why it should be brought in *àpropos* of vegetables. It is the staunch conservatism, where food or cooking is concerned, of the working classes of England. In politics they are very often to a man, nay, even to a woman, advanced Liberals, to say the least of it. They are much more ready to advocate and adopt sweeping changes in things of which, after all, they cannot know a great deal; but they distrust anyone who suggests that they could improve the matters which lie close around them, and with which they are at least familiar. "My ould grandmother did it that way, and she lived till ninety," is an unanswerable argument against making the scrap of meat into a *pot-au-feu*, and adding vegetables and meat to it, instead of frizzling and burning the same scanty portion of meat in a greasy frying-pan over a smoky fire. I feel persuaded, therefore, that the great reform in cooking and economic management of our food-material must *begin* in the classes above the working man. When he sees and learns by experience that an ounce of meat, properly dressed, will go further in actual nourishment and strength-imparting qualities than two ounces heated in his old barbarous method, he may perhaps be induced to consent to his "missis" or the "gals" being "learned" how to cook. My own private hope— and I would almost say expectation—is, that an increase in the artisan's or the working man's comfort at home,—such comfort as better cooked food and more of it must surely bring,—will lead to his wages finding their way oftener into the butcher's shop than the public-house. A well-fed man is very seldom a drunkard; and it may be that in the spread and development of an attempt at culinary reform, two birds may, all unconsciously, be killed with one stone. In improving cottage comforts we may perhaps strike a great blow (with our frying-pans and soup-kettles!) at the shining glasses and quart pots of the gin-palace. God grant that it be so!

LESSON VII.

MODES OF PREPARING BROTH OR SOUP FROM BEEF.

THE reason I have placed this subject in a separate lesson is because of its enormous importance in the sick-room. More delicate children are reared into health and strength, and more lives are saved, by good beef-tea than most of us have any idea of. This is the more extraordinary when we remember that even the strongest and best beef-tea contains an almost infinitesimal amount of actual nourishment. So that it is not to its capacity for supplying to the wasted and feeble human frame either strength or nourishment that we must attribute its wonderful efficacy. If the strongest beef-tea be analysed, the meat would be found to have lost in the process of turning into liquid nearly all its albumen, fibrine, and caseine. In other words, it would have parted with its most important constituents; and we might suppose it therefore to be valueless to the human system. But Experience steps in where Chemistry stops and shakes her head, and Experience declares that well-made beef-tea possesses a reparative power on a weakened digestion which nothing else in the world except milk can come near. It may not actually contain all the elements of nourishment within itself, as milk does, but it is a wonderful assimilator. It soothes and repairs and collects the enfeebled organs and juices, and enables them to return to their proper functions. Therefore we say that beef-tea is nourishing, when it is not in the least nourishing in itself, but it has the power of making ready for other substances to nourish.

Although every sort of meat can be made into soup or broth, bee makes the best and wholesomest. For one reason of this we must search in the fibrine, which holds more red juice than that of any other meat, and it is this red juice which we particularly want. Everybody knows that the leanest meat is the best for soup-making; the least particle of fat is out of place in broth or soups, and indeed renders it absolutely unwholesome as well as nauseous.

In many emergencies beef-tea has to be prepared at almost a moment's notice, and then I would recommend that the meat be as thoroughly freed from fat as possible, chopped finely, and soaked in its own weight of cold water for ten minutes or so. Then heat it slowly to boiling-point, let it boil for two or three minutes, and you will have a strong and delicious beef-tea, better than can be obtained by boiling in the ordinary way for many hours. Another method is to place the finely-chopped meat in a large, clean jam-pot, with a little water and a pinch of salt. The mouth of the vessel should be closed by means of a tightly-tied bladder or a thick paste all over it, as if it were a meat-pudding, and placed in a saucepan half full of cold water. The saucepan should then be covered with its own lid and set upon or by the side of the fire to simmer slowly. If there be no time to let the beef-tea or essence in the jam-pot get cold, it must be skimmed as clearly as possible, and any extra globules of fat floating on the surface removed by a careful application of white blotting paper. Some people do not add any water at all to the cut-up beef, under the impression that the essence must be stronger without the addition. But my individual experience teaches me that whereas the difference in nutritive value is very slight, sick people do not like the beef-tea thus prepared, and will not take it so readily as when it has been made after the following manner. It is necessary, however, to state that the process I am now going to describe *cannot* be hurried, and that it is therefore imperative to have one day's notice when beef-tea made in this way is required.

Take two or three pounds of the leanest beef to be procured, add one quart of water, and two shank bones of mutton, which bones should be well washed before using. A pinch of salt, and another pinch of grated lemon-peel, or a tiny bit of the peel itself, are all I should add, for a sick person's throat is generally too tender for pepper, and his palate too delicate for anything like flavouring or sauces. The lean meat and shank bones are to be put into a saucepan, whose white enamelled lining should be daintily and scrupulously clean, and the saucepan, with its lid fitting very close indeed, set by the side of a moderately good fire to simmer slowly the whole day long. It must never approach boiling, and yet the action of fire upon its contents should be decided, though gentle. At the last moment before shutting up for the night, strain the soup through a fine hair sieve into a clean basin, and in the morning you should find, beneath a preserving scum of fat, about a pint of clear, solid, beef jelly, which can either be eaten cold, or warmed, without the addition of one drop of water, into a delicious *clean*-tasting cup of beef-tea. In cold weather double the quantity may be made, but in that case it should be poured into *two* basins, and the fat left to hermetically seal the second basin until it be wanted in its turn for use. In hot weather the beef-tea should be prepared fresh *every* day for the next day's consumption. I have seen beef-tea rendered perfectly colourless and white by repeated strainings through fine muslin sieves, but I do not know that this is any particular advantage.

In some cases, such as the terrible state of the intestines after typhoid fever, beef-tea is no use as a reparative agent when prepared after the above fashion. The meat should then not be cooked at all, only cut up as lean and fresh and full of red juice as possible, and soaked for ten or twelve hours in a small quantity of *cold* water. This will give a liquid which has never been submitted to the action of fire, and which looks and tastes like the gravy of under-done meat, but it is of the highest reparative value to the lacerated stomach. A judicious nurse will take care that her patient never *sees* this sort of beef-tea until he has learned to drink it freely, which he will do if not at first disgusted by the sight of the clear red fluid.

I have dwelt thus minutely on the value and process of making beef-tea because I believe it to be the strongest resource of the culinary art in sickness; but the proper preparation of soup is of great importance in all households. It is at once an economical, wholesome and savoury form of nourishing food; yet, to many a *plain* cook, soup, unless she has costly materials bought expressly for its manufacture, merely means greasy hot water flavoured by a *soupçon* of plate-washing! No soup should be used the same day it is made, on account of the impossibility of removing all the scum and fat. But, supposing that a scrag end of mutton, or the trimmings of cutlets, or bones with a fair amount of meat left on, should have been simmering gently all the preceding day, and allowed to get cold at night, so that the layer of fat (which can be used for other purposes) is easily removed, then we should proceed this way, always imagining it is wanted for the use of a poor and economical family. To the clear, fat-free soup, add half a tea-cupful of well-washed pearl barley or rice—and we must remember that the inferior and cheaper kind of rice does just as well as the best for this purpose—a few cleaned and cut-up vegetables, a little onion, pepper and salt, a sprig or two of herbs tied together, a little pea-meal, any cold potatoes left from yesterday's dinner, and the whole allowed to simmer together, without removing the remains of the meat and bones, until it be wanted, great care being taken that it should not boil away. The result of this simmering *ought* to be a nice, warm, comforting, *clean*-tasting basin of broth, very different to the weak, greasy liquid which results from a hastier preparation. It is a very common mistake with all cooks, except the very best, to put too much water in the first instance to their materials for soup, **and** so produce a good deal of weak, tasteless **meat-tea,** instead of a smaller quantity of strong, good soup. English people do not use macaroni half so freely as they might, for, apart from its nutritive value as offering such a pure form of wheaten flour, it is exceedingly cheap. Boiled with ever so little soup made in the way just described (before the addition of the rice or vegetables), it would form an excellent and wholesome change to the smallest bill of fare.

All cooks prefer beef to anything else for making soup, but a very nourishing and delicate broth can be made from two parts of veal and one part of lean beef, or from chicken or rabbit, though the latter is not advisable for sick people. Everyone knows the value of good, fat-cleared mutton broth such as I have just described, but there is a good deal of truth in the instinct which leads the sick person to prefer beef-tea, and the healthy labouring man to buy a couple of pounds of beef instead of double the quantity of any other meat. Beef contains most iron, which in the state of oxide is one of the chief constituents of the blood : and we must bear in mind that the nutriment of all carnivorous animals is derived from the blood originally. A diet, therefore, to be strengthening, must contain a certain amount of iron, and we do not obtain this so readily from any other meat as from beef.

LESSON VIII.

FUEL AND FIRE.

THE object of cooking is to render the flesh of animals and vegetable substances easier of mastication, and therefore easier of digestion. How this object is carried out in most English households let each declare for himself. And yet there is nothing in the world so simple and so certain in its effects as the action of fire upon food, if only we can learn to apply and to regulate that action according to certain laws. I propose therefore to devote a short lesson to each of the simplest processes of cooking.

But before doing so I may be permitted here to say a word or two about the management of the kitchen fire. Few ladies, or even those servants whose duties lie entirely upstairs, and who see a bright or blazing fire every time they go into the kitchen, can have any idea how difficult a thing it is to keep up a good fire all day. When I say a "good fire," I mean a good *cooking* fire—a clear, bright fire, which, without being a roaring furnace, shall yet be equal to any emergency. It can only be managed by constant small additions of coal, unless a great deal of cooking is imminent, and then of course more fuel must be added each time. But a really good cook will so contrive as to have a small, bright fire all day long, even when she is not actually cooking. Whenever I hear that a bit of bread cannot be toasted, or a cup of soup warmed, because the fire has "just been made up," I know what has happened. The cook has allowed the fire to burn down to the last bar of the grate, and then she has emptied half a coal-scuttle on the few live embers. For about two hours, therefore, it is useless to expect any cooking from *that* fire, and it will be fortunate if no sudden call be made for its services. Now, if the cook had watched her fire, and had kept it supplied from time to time with small portions of coal, this emergency would never have arisen. She could screw up her fireplace to very small dimensions and yet keep an excellent fire, fit for any unexpected demand. It is doubtful whether, when she acts on the momentary impulse of trying to make up for lost time, a cook has any idea of the mischief she does. Letting the kitchen fire burn low and then flinging on coals, is not only an inconvenient, but it is a recklessly extravagant proceeding. The fire and fireplace have become thoroughly chilled, and the fresh fuel evaporates almost entirely in the form of smoke for a long time before the remainder is in a state to use for cooking.

If this rule of preventing waste by constantly adding small portions of fuel were better understood and acted upon, cooks would not have such a bitter prejudice against the use of coke. It is, of course, absolutely valueless to a half-extinguished fire, especially when, instead of being put on in small quantities, it is flung on in shovelfuls. But to an already clear, well-established fire, nothing is so satisfactory or economical an addition as a few lumps of coke judiciously put on. If frying or broiling is to be done, the fire *cannot* be too clear, and coke, if it be properly managed, will give the clearest fire in the world, but then it requires a certain amount of intelligence and willingness on the part of the cook to use it to advantage. When I use the word cook, I do not mean only a regular servant, but any young woman who is acting, for perhaps the first time in her life, the part of cook in her husband's, or father's, or brother's house. She will find her culinary labours much simplified if she keeps the needs of the kitchen fire always before her mind. I don't mean to say that such a one may not what is called "make up" her fire, and leave it untouched between breakfast and dinner, and dinner and tea, because the chances are a hundred to one she will not need it, and her duties probably call her elsewhere; but a cook in a house where there is a family, and perhaps sickness, or even very young children, ought never for one moment to forget or neglect her fire all through the day.

I *could* give her scientific reasons about radiation, and use many long words to prove to her why, if she keeps her grate well blacked and polished, she will find her fire burns better and gives out more heat, but I prefer to appeal to everybody's experience and common sense if such warmth and brilliancy be not the result of a beautifully clean and shining fireplace.

To Sir Benjamin Thomson (an English knight and an American by birth, but better known to us by his Bavarian title of Count Rumford) we owe perhaps more improvement in the economical management of fuel and the construction of stoves and fireplaces, with due regard to that economy, than to anyone else in modern times. He was induced to turn his attention to the subject by the scarcity of fuel on the Continent, and his ideas naturally expanded and enlarged themselves by constant practice. At last he succeeded in inventing a method of heating houses and of cooking food which did not require much more than half the usual amount of fuel, and this economy in firing became such a mania with him that the joke of the day used to be that his highest ambition was to be able to cook his own dinner by means of his neighbour's smoke.

However that may have been, it is very certain that to Count Rumford we owe a great increase of our knowledge on such subjects, and the reason I mention him particularly in this place is that he never seemed to weary of insisting on the necessity of a well-kept brightly-blacked fireplace to the due economy of the fuel used in it. He explained incessantly how that kind of heat which is absorbed by either black or white surfaces is totally devoid of light, and may almost be considered as pure, radiant heat. So that the first point to be taught, in ever so humble a kitchen, is that the fireplace should be exquisitely clean, besides well and brightly blacked, in order to give the fuel which will be used in it a fair chance of giving out, by radiation, every particle of its latent heat.

The next thing to be considered is the division and arrangement of that fuel, beginning from even the starting-point of lighting the fire. A careful housewife—careful either on her own account or her mistress's—will only use half as much wood or shavings to start her fire with as a thriftless one, because she will take trouble to learn that there is a scientific but perfectly simple mode of laying and lighting a fire. She will be told in theory, and prove for herself by practice, that she must thoroughly clear out her grate, clean and brighten it up to the highest pitch, and then place in it whatever is her lightest material, her paper, or dry grass, or shavings, whatever she has at her command. Next come the slender twigs or dried sprays of heather of the country, or the neatly-cut firewood of the town. Unless all this is thoroughly dried over-night, it will be worse than useless, and it is in attention to details of this sort that true economy consists. A damp bundle of wood or twigs will smoulder, and be consumed without making any appreciable difference in the state of the fire, whereas half the quantity, when thoroughly dry, will start a satisfactory blaze in a few minutes. Then should the cinders be thoroughly and carefully sifted; and now-a-days I have no hesitation in saying this is as imperatively necessary in a palace as in a cottage, on account of the increased price of coal. No cinders should be relegated to the dusthole at all, for everything, except actual dust or the hard flakes (called clinkers) left by coke, can be used. The largest cinders may be laid lightly on the logs of the blazing sticks, the smaller ones being thrown up, later, at the back. Cinders are the best material in the world for starting a fire, and even small lumps of coal should only be sparingly used at first. Above all, a beginner should be taught that her fire will *never* light or burn up if she does not take care to establish a free circulation of air beneath. I am, of course, speaking of ordinary open fireplaces. Stoves and other patent fireplaces are generally constructed on entirely different principles, and require special instruction for the management of their fuel, but this is easily obtained from the person who fixes them.

Taking it for granted, then, that our ideal cook thoroughly understands how to light her fire, and is impressed with a due sense of the importance of a well-blacked shining kitchen-range, or humbler tiny fireplace—the rule is the same everywhere—and that she is one of those capable people who would disdain to shelter themselves behind the excuse of an ill-tempered chimney or a "bad draught," we will presently proceed to see what she should cook upon her fire.

PART III.

THE PRINCIPLES OF DIET AND A FEW CHEAP AND EASY RECIPES.

PART III.

REMARKS.

THE first principle of diet is that the stomach should not be asked to receive more than it can digest; and the second, that the food should be suitable to each person's digestion. We are very tyrannical to our stomachs, and they, in their turn, generally retaliate upon us sooner or later. If a certain form of diet agrees with one individual, it is no absolute rule that it should suit our neighbour; but we too often insist on feeding others according to what we imagine agrees with ourselves. Especially is this the case with children's diet, and few grown-up people make allowance for the healthy appetite of girls or boys who are still growing, or understand how much food-material the rapidly-expanding frame requires.

My own firm conviction is that no schoolboy ever gets as much nourishing food as he requires, and that that is the secret why boys of fourteen or fifteen years old scarcely ever look anything but thin and pinched. The general remark is, "Oh, they are growing so fast!" So they are, and that is the exact reason why their food should be particularly nourishing, more so than at any other time of their lives. Instead of that, an

English schoolboy gets *two* slops and only *one* nourishing meal a day, during the years of his life when he requires the greatest amount of nutritive food. Think of the actual force-producers contained in a schoolboy's breakfast and tea (or supper), and think of the amount of exercise his restless young limbs will take or have taken in the course of the day. After a game of football or cricket, or a paper-chase, a boy sits down generally—I might almost say invariably—to a meal of weak tea, skim milk, bread, and perhaps cheese or a little butter. I am not, of course, speaking of cheap schools. When a person undertakes to feed and teach and board a boy for a sum between 20*l*. and 50*l*., or even more, it is well-nigh impossible, at the present scale of prices, to give him better, or even as good food as what I have described; but it does appear to me a shame that at the more expensive schools to which boys are sent by parents of fairly good means, the scale of diet should be kept so low, and the proportion of really nutritive food so small. Perhaps the only exceptions to this rule are to be found in the liberal tables of some of our best public schools, but even there the boys, without being absolutely starved, do not get enough to eat, and two meals out of the three will probably contain insufficient nourishment. In girls' schools, I fancy, this evil is still more decided, and a poor diet whilst a child is growing rapidly is the root of delicate constitutions, feeble frames, and general "breaking down" at the outset of life.

There should also be the greatest imaginable difference in diet between different classes of workers; for although a certain section of the community monopolizes to itself the honourable title of *the* "Working Class," the term embraces many more thousands than the labouring man imagines. The popular idea, for instance, among the poor and ignorant masses who work for their daily bread, is that the Lady who rules over this country leads a blissful life of idleness, seated on her throne all day, orb and sceptre in hand, and gazing placidly before her into space. Now, I believe it to be a fact that few people in all Her wide dominions work really harder, in every sense of the word, than our dear and good Queen. At the head of the workers her Majesty may well claim to take her place, and then will come a crowd of men and women who wear good clothes and live in fine, or at all events decent, houses, and yet work absolutely harder, all the year round, than any day labourer in the Midland Counties.

The diet for work of this nature must necessarily be very different to that required by the man who exercises his muscles in the open air, and whose appetite and digestion possess far larger capacities of receiving and assimilating food than those of the poor brain worker who uses up his life-power at a much quicker rate. The absence of fresh air, and the want therefore of constantly renewed supplies of oxygen to the blood through the lungs, prevent the man who works indoors with his head or his hands from feeling so hungry, yet the exhaustion of his nervous system demands as urgently that it should be renewed by means of food. At the same time the digestion of such a one is weaker, and cannot manage gross substances. For these workers, then, a diet where the cooking is so perfect, however simple it may be, that there shall be as little strain as possible thrown upon the gastric juices, is of the first importance. To brain-workers albumen is even more necessary than fibrine, and raw eggs afford this in its purest form. There is a popular fallacy that eggs beaten up in milk are rendered doubly nourishing, but if the egg be fresh and good the combination is rather more fitted to hinder than to promote digestion. It would be better to beat the egg up in a little brandy or wine, and wine is the best. Fibrine, in the form of meat, should be sparingly used by those who live by their brains, and the meat should be of the best quality, and always very well and delicately cooked. Fish supplies most easily the phosphorus which is needed by such a system, and good pure milk and cream are also very essential articles of diet.

But to the man who exercises his muscles in the open air a very different regimen must be prescribed. The labourer instinctively stops the gaps between his scanty meals with cheese, which is the best thing for him, and he enriches his poor diet of potatoes with bacon. Some day, when his wife has learned how to make the most of every scrap of meat, he ought to be able to vary his food with a good drop of warm nourishing broth. If only he could be persuaded to diminish his beer and increase his allowance of meat, he would find himself in a far better condition for work.

The diet of our soldiers, and even of our sailors, appears to me—in spite of tables showing the proportions of flesh-formers and starch, of gluten, and heaven knows what, swallowed daily by every soldier—to be really insufficient for a healthy man with a good appetite. They may be supplied with food enough to prevent anything like actual starvation, and even to keep them in some sort of condition, but I question whether a British soldier ever knows what it is to feel thoroughly satisfied after his meals for one whole day. It is just possible, is it not, that the men would be easier kept away from the canteen if they had as much as they could eat? Tables of food-proportions are very well in their way, but I know that I have seen working men in New Zealand, and growing boys of eighteen and twenty years old in colonies where meat was cheap, consume fibrine—or, in other words, eat plain roast meat—in quantities which would soon leave the most liberal military dietary several pounds behind.

It is not at all certain that, in spite of danger and discomforts, our soldiers do not really fare better abroad, or in time of war, than at home in peace. In the face of a national excitement we are not so very particular as to the number of ounces of meat to be dealt out to the men who have to stand between us and ruin, so the soldier has then a better chance of occasionally getting as much as he can eat. If he could cook his own food, he would be still better off; and anyone who saw those good-looking German soldiers cooking their rations in the little tent behind the School of Cookery last summer, must remember how deftly they set about their preparations, and how savoury was the result of a pea-sausage and a bone or two. No doubt every year brings its improvements in these matters, and if a soldier who fought under Marlborough could see the rations and barrack accommodation of his modern brethren-in-arms, he would indeed think they had nothing to complain of in the way of food and shelter. But still there is ample room for improvement, and I would endorse the suggestion often made before, that the British soldier be taught to cook, and to make the most of his rations by such cooking. Each man might take it in turn to try his hand over the fire, and there might be some regimental emulation in the form of small prizes for clever contrivances to vary the food, and so forth.

I am aware that the food is not nearly so monotonous as it used to be a short time since, when all the meat eaten by soldiers was invariably boiled; but still I question whether the mess dinner of the rank and file is anything like so savoury and palatable as the dinner to be had a few years ago in Paris, at one Madame Roland's, near the Marché des Innocents. For twopence she gave you cabbage soup with a slice of *bouilli* (beef) in it, a large piece of excellent bread, and a glass of wine, which it must be admitted, however, was rather thin. Some 600 workmen used to throng daily round her table in a shed, and yet she calculated that she gained a farthing by each guest. In Glasgow, Manchester, and elsewhere, similar public dining places have been established on the cheapest possible scale, and found to answer very well; but although a workman may be able to get a fairly good and nutritive dinner at such an institution, it is not the less necessary that his wife should know how to cook his food decently for him at home.

LESSON IX.

BOILING AND STEWING.

THERE is all the difference in the world between boiling meat which is to be eaten, and meat whose juices are to be extracted in the form of soup. If the meat is required as nourishment, of course you want the juices kept in. To do this it is necessary to plunge it into boiling water, which will cause the albumen in the meat to coagulate suddenly, and act as a plug or stopper to all the tubes of the meat, so that the nourishment will be tightly kept in. The temperature of the water should be kept at boiling-point for five minutes, and then as much cold water must be added as will reduce the temperature to 165°. If the whole be kept at this temperature for some hours, you have all the conditions united which give to the flesh the quality best adapted for its use as food. The juices are kept in the meat, and instead of being called upon to consume an insipid mass of indigestible fibres, we have a tender piece of meat, from which, when cut, the imprisoned juices run freely. If the meat be allowed to remain in the boiling water without the addition of any cold to it, it becomes in a short time altogether cooked, but it will be as hard as iron, and utterly indigestible, and therefore unwholesome.

If soup is to be made out of meat, then it stands to reason we want all the juices which we can possibly extract from the meat to mix with the water. Therefore the meat should be put into *cold* water, with a little salt and a few vegetables (if in a poor family a few crusts of bread may be added at the last minute), and allowed to simmer as long as possible. It is undoubtedly the most economical form of nourishment which exists, and it is an absurd prejudice to suppose that the same amount of meat is invariably more valuable to the human system if it be frizzled in a greasy frying-pan, so that it becomes burnt outside but remains raw within, and eaten in this state as " good solid food," dear to the heart (but surely not to the stomach) of a true Englishman. In the first place, even a pound of meat will only feed one person in a solid form, whereas, if to exactly the same weight of meat be added a pint of cold water, a few vegetables, or even herbs, a couple of potatoes, a bone or two, a scrap of bacon, an onion—almost anything which comes handy—we have at once the *pot-au-feu* of the French peasant, and produce a warm, savoury, wholesome meal for two or three persons. It may be as well to mention that the scum which rises on the top of the water whilst meat is boiling is *always* useless and unwholesome, and should be got rid of as completely as possible. The way to help this scum to rise, so as to be able to get rid of it, is to keep pouring in a little cold water from time to time. This will always have the effect of sending up some of the obnoxious substance to the top, from whence it should speedily be removed.

Stewing occupies a sort of middle position between roasting and boiling, and must be carefully attended to, if the meat is not to be hardened instead of softened by the process. It is desirable to dip meat into boiling water for stewing as well as boiling, unless indeed it should have been soaked before. What, for instance, makes hashed mutton a byword of nastiness? Because an ignorant cook plunges her chunks of cold meat into a greasy gravy when it is at boiling-point, thereby thoroughly and hopelessly hardening the meat, and then serves up the mess with large pieces of half-toasted bread. Now, is this way more extravagant? I can answer for its being more palatable. Make a nice little gravy of any cold stock —and a good cook will *always* have a small basin or cup full of stock by her—add an onion finely shredded and fried, a little pepper and salt, and, if it is to be had, a tea-spoonful of ketchup. Let the mixture come to boiling-point, without boiling over, and strain it into another saucepan. If you have only one saucepan, strain it into a basin, quickly clean out your saucepan, and pour the gravy back into it, setting it aside to let it get nearly quite cold. *Then*, and not until then, lay in thinly-cut, small slices of the cold meat, and let the gravy and the meat warm thoroughly and gradually together, *without* boiling, but don't allow it to stew too long. Whilst it is getting ready, have the frying-pan ready with a little boiling fat (not that which fish has been fried in, remember), and put into it some small, thin, three-cornered pieces of bread, which will quickly fry into a crisp toast. Serve these round the hash, which, by the way, should not be swamped in gravy, and I can answer that a certain cockney millionaire friend of mine will no longer issue this solemn warning to his family: "Never eat 'ashes away from 'ome."

But to return to stewing. If it be properly understood and practised, stewed meat makes a very agreeable and palatable change from the monotonous boiling and roasting which alternate on the middle-class daily bill of fare. A shoulder of mutton stewed, Indian fashion, with a handful of well-washed rice, a few Sultana raisins, half a dozen cloves, and a teaspoonful of currie powder to flavour it, makes an agreeable change. Some meats are far more wholesome also when stewed than when roast; as veal, for instance, and many kinds of fish. Eels are invariably more wholesome stewed than boiled—though *all* fish is wholesomer boiled than fried, for stewing is a more gradual process than boiling, and the fat is more surely got rid of. If it should ever be necessary to cook a beefsteak which has not yet had time to become tender by keeping, then, for the sake of the digestion of the family, it would be better to stew it, and this is the way it should be done.

The meat should first be cut into convenient, but large-sized pieces (all the fat having been removed) and lightly fried on both sides in butter or clarified dripping. This will make it of a nice brown colour, and prevent the pale flabby appearance it would otherwise present. Then get a saucepan and put the meat into it, with a little sliced onion, turnips and carrots (which are also improved by being half-fried first), pepper and salt, and a tea-spoonful of any sauce you prefer. If there is any stock, add it, but if not, put in about half a pint of water, and let it all simmer very gently for two or three hours. At the last moment skim it well, for it is odious if it be greasy; stir in a few pinches of flour to thicken the gravy, and let it all boil up together for a couple of minutes before serving. Some people are very fond of fat with all their food, though they should bear in mind that fat affords no nourishment whatever to the human body. It merely goes to make fat. A stout person should therefore not eat much fat, and a thin one should. The function of fat, as we all know, is like starch or sugar, to keep up the heat of the animal, and a certain proportion is even present in healthy animal muscle;

so it does not do to buy lean meat, although all the fat on the joint need not be sent up to table. However, it is necessary to serve a certain portion of fat with stewed steak, but do not let it stew *with* the meat, for it will only melt and rise to the surface in the scum which has to be so carefully removed. Rather keep the fat till the last moment, cut it into little pieces a couple of inches long, and put it by itself in the frying-pan or on a gridiron for a minute or two just to cook it, and serve it in golden-brown nodules on the top of the stewed meat.

All nice cooking—be its materials ever so simple— is more or less troublesome; but I have always found (and the experience of others bears out my own) that bad cooks will take quite as much trouble to spoil food. It is therefore a great pity that when a woman is conscious of her own deficiencies and is anxious and willing to improve by learning, she should not have the opportunity of doing so. But unfortunately cooking is not to be learned from a book, nor from a lecture. It is an art in which practical experience, supplementing theoretical information, alone can be of any use. It is doubtless a great advantage to intelligent beginners to have the why and wherefore of everything explained to them either by voice or page, but it is equally necessary for them to see with their own eyes and try with their own hands the result of these instructions, for half-an-hour's practice is worth a week's theorizing, in cooking as well as in other things.

LESSON X.

BAKING, ROASTING, AND FRYING.

THE same principle which has been advocated in boiling holds good with regard to roasting. If you wish to retain all the juices in the meat, place it close to the fire for five minutes *at first*, and then remove it to a greater distance until the last five minutes, when it should be brought near the fire again. It is possible, by this method, to roast a joint thoroughly, so that it shall be perfectly well cooked, and yet, when carved, the imprisoned juices shall flow out readily. All meat ought to be well floured and sprinkled with a pinch or two of salt before putting it to the fire, and it should be kept constantly basted with clear dripping. Some things, such as hare, are better basted with milk; and poultry, or any very small joint, is much improved by being covered with lard or oiled paper. Instead of larding game or poultry, it is often preferable to *bard* it, *i.e.* to cover the breast with a thin slice of fat bacon, which may be served up with it as with quails.

We must remember that the object in cooking is to present meat, and indeed all food, to the palate in an agreeable form without changing its composition more than we can help, or losing its nutritive value. Raw meat, quite apart from other objections, is so tough that it would be impossible to masticate or digest enough of it to satisfy hunger, whereas the application of heat is intended to force the juices to expand, thus separating the fibres and making mastication easy and pleasant.

The loss of weight in roasting, especially if the joint be a fat one, is very considerable. As much as 4 lb. 4 oz. have been lost in roasting a joint of 15 lbs. weight in the ordinary manner. Although meat actually loses more of its weight by roasting than by boiling, yet, if no account be taken of the matters extracted, it contains, when roasted, a larger proportion of nutritive elements than the larger mass of boiled meat, and in a given weight is more nutritious. Meat is often baked, and though this method may be harmless and agreeable as a change, it is not such a wholesome form of cooking as roasting.

The primitive manner of baking meat is the only one which ensures it from becoming dry and tasteless, namely, to enclose it in a crust of some sort. The gipsies to this day bake their meat and poultry—we will not inquire how this latter item is added to the bill of fare—in a sort of mud mould or case, covering up feathers and all; and the Indians and Maoris generally cook in the same way. A fowl, or a piece of meat of any sort, is delicious when enclosed in a flour-and-water case—dough, in fact—and baked in the embers of a camp fire. If the meat were put in the fire without this protection, it would simply get burnt.

Frying is the simplest, the commonest, and, if properly done, the wholesomest form of cooking food, but it is perhaps the least understood, and more often results in burning the outside of the meat whilst the inside is left raw. To begin with, a clear, smokeless fire is indispensable for frying, and it is equally necessary to have a perfectly clean frying-pan. Of course the best oil, or the best fresh butter, would offer the most perfect conditions of the fat in which anything should be fried; but good, pure, clear fat, and clarified dripping, make capital substitutes. Cold meat is excellent when lightly fried and served up with yesterday's vegetables and potatoes (also cut up and fried), but the excellence depends entirely on the delicate yet savoury flavouring, the clearness of the fire, and the goodness of the fat in which the frying process is carried on. It is also very important that the fat should be actually boiling. Here again we are met by prejudice, for ninety-nine cooks out of a hundred will allege that they are "respectable women" when asked to use a frimometer or a thermometer, and prefer to go on ascertaining the temperature of their fat by guesswork or by means of a sprig of parsley. It is more economical to roast the flesh of young animals, such as lamb, chicken, veal, or pork, because such flesh contains an undue proportion of albumen and gelatine in the tissues, and these substances will to a great extent be lost in the boiling.

If I had to cook a dish of cutlets and potatoes, or a tender rump-steak and potatoes, this is the way I should do it, or, to speak quite truthfully, these are the directions I should give for its being done. First, I must say that whenever it is practicable to use a gridiron in the place of a frying-pan, and to broil meat instead of frying, it should be done. But, at the same time, I *have* tasted such excellent cutlets served out of a frying-pan, that it shows it is not an invariable rule. It is the attention to small details which makes all the difference in nice cooking, and if persons thoroughly understand the value of these important trifles, they learn to do the thing always that way, and so it becomes no more trouble to them than is the slatternly method which results in grease and cinders, heartburn and disgust. Well, then, let us imagine that we are rich enough to possess a frying-pan *and* a gridiron, and that our fire, however small, is clear and bright, without a film of smoke, for it is of no use trying to fry or broil unless the fire is in a proper condition. In spite of what has been said in a former place about cooking potatoes in their skins, potatoes for frying must needs be peeled, well washed, and cut rapidly up with a sharp knife into thin slices. Again, they should be thrown into a basin of water for a moment, and then laid on a clean cloth, slice by slice, to be thoroughly dried. All this time the nice, clear fat should have been melting on the fire, and when it is actually boiling throw in the potatoes, keeping the frying-pan frequently moving so that they shall not stick to its bottom. A couple or three minutes ought to crisp them to a beautiful golden brown colour;

then skim them swiftly out of the boiling fat, throw them into a large, fine wire sieve (which would be all the better for having been warmed to receive them), sprinkle a pinch of salt over them, and turn them into a very hot dish, every particle of fat having been left behind in the sieve. Although the potatoes have been mentioned first, the meat should really have preceded them in the order of cooking, as it is the easiest to keep hot. If you are going to have cutlets, trim them from the best end of a neck of mutton very neatly. There is no occasion to throw away the scraps; they should either go into the stockpot, or, if strict economy be necessary, they may afterwards be made into a pudding or pie. The chine-bone must be sawn off, and the seven or eight chops (which are all you will be able to get off a moderate-sized neck of mutton) neatly pared, and only about an inch of bare bone left to each cutlet for a handle. The cutlets should then be sprinkled with a little salt and pepper, and laid for a moment in a dish of oil; then put them on the gridiron, or into the frying-pan, but in this latter case add a little more oil, and broil or fry them for six or seven minutes. They ought by that time to be nicely done, and should be served hot. Beefsteak can be cooked exactly in the same way, only from its larger size the gridiron is more strictly indispensable. A frying-pan is a very serviceable implement in the hands of a skilful manager. I trust she will make it a point of keeping it scrupulously clean, and then she can serve up the cold vegetables left from yesterday in this fashion at a moment's notice. Melt a little fat or butter in your frying-pan, shred an onion into it with a spoonful of chopped parsley, a little salt and pepper, and a sprig of any savoury herb or bit of lemon-peel which comes handy. Then cut up the vegetables—cabbage, turnips, carrots, and so forth—into small pieces, and fry the whole, lightly tossing the contents of your frying-pan all the time, so that they may not get into a burnt fat-soaked mass. On a sudden call for a late supper, such a dish as this forms a capital addition to the cold meat or fried bacon and eggs.

Of all the uses, however, to which a housewife turns her frying-pan, I suppose an omelet is the least in demand, and yet it is at once the cheapest and easiest way in the world to cook eggs with other things. All it requires is vigilance and knack. Don't *over*-beat your eggs, just whisk them up (three are quite enough for a manageable omelet), whites and all, lightly and swiftly, beat in with them a pinch of salt, a little pepper, some finely-chopped parsley, or a teaspoonful of grated cheese, or shredded bacon, or even shredded fish ; almost anything mixes well in an omelet, provided it is cut fine enough. Have the frying-pan ready on the fire with butter enough in it to fairly cover its surface when melted, which it should do without browning. Into this clear liquid butter pour the contents of your basin (your eggs, &c.), holding the frying-pan with the left hand, and gently stirring the mixture with a wooden spoon in the other. The omelet will set almost immediately, and then the stirring should be discontinued, and the gentle shaking carried on *incessantly:* the edges being lightly turned up with the wooden spoon every now and then. If you turn your head, or cease shaking for a moment, the omelet will be spoiled. Four minutes should be quite enough to cook the inside thoroughly, and yet leave the outside of a rich, yellowish-brown colour, but the time required to attain this result will entirely depend on the fire. Too fierce a fire will burn the omelet before it has had time to set or become thoroughly cooked, and yet a clear brisk fire is necessary. As soon as it begins to assume the shape of a small plate and the colour of a golden pippin, take your wooden spoon once more and dexterously double it over, serve it in an exceedingly hot dish, and eat it whilst it is still sputtering and frothing. The only things requisite in an omelet are, presence of mind and promptness of action. Timidity and hesitation have ruined many an omelet, and it is better to practise as often as may be necessary, before serving up a failure.

In fritters, the yolks of the eggs and the dissolved butter are beaten into a batter, and the slices of fruit, previously dipped in finely-powdered sugar, dropped into the mixture, to which, by the way, the well-whisked whites of the eggs must be added at the last moment. Then the slices of fruit, with the batter adhering to them, may be placed in the buttered frying-pan for a moment or two just to get lightly cooked, and the pan should be kept well shaken during the process.

LESSON XI.

BACON.

AMERICAN bacon is considerably lower in price than English bacon, but it shrinks more when boiled, and you can get a larger number of slices from a given weight of English bacon than can be obtained from the other. Pork is the great stand-by of the poor man's dietary, by reason of its strong flavour as well as its low price, and the relish it affords to monotonous and insipid fare. The dripping from fried bacon is often preferred by children to the rancid stuff sold as butter to the poor; and in any case the fat from bacon is more palatable with cabbage or potatoes than the suet of either beef or mutton could possibly be. It is easier to carry when cold into the fields; and another great advantage of bacon is that it requires less fire to cook it, and fewer utensils. From a scientific point of view, a diet in which bacon is the principal meat, needs to be largely supplemented by milk and other highly nitrogenous food, for it contains very little nitrogen itself, and we know that nitrogen is of great importance to the blood. Bacon supplies a fair amount of carbon, and does not therefore require the aid of bread. With the addition of a little pea-meal, the liquor in which bacon has been boiled makes a good soup, and it would be improved both in flavour and nutritive value by a few potatoes and an onion being boiled in it.

But as a general rule, however valuable the pig may be in an economical sense, it is quite certain that pork is less wholesome than almost any other meat. For the reasons why this should be so, we must go in the first place to the habits and ways of the animal itself, its absence of any guiding instinct about food—for quantity, not quality, appears to be the first principle of a pig's diet—and the motionless life it leads. Pigs which are turned out in a field run about too much to grow fat, and therefore, if it be necessary to use the animal for food, it is speedily relegated to its sty. There it never does anything except sleep and eat, and this want of exercise tells not only on the inordinate growth of fat which is laid up outside the body, but upon the muscles and fibres of the flesh, which become hard and indigestible. The pig stores up in its body three times more of its food than the ox, and from its large proportion of fat is not of equal value with beef or mutton in nourishing the system of those who need to make much muscular exertion. The leg of pork is the part of the body which, if deprived of its large proportion of fat, approaches the most nearly to the nourishing elements of beef or mutton. How ever, I do not for a moment expect that any scientific theories for or against pork will have any ill effect on the keeping of pigs or the curing of bacon. Happy is the family which can keep a pig; therefore, what does it matter whether it be a "highly nitrogenous food" or not? Piggy pays the rent, and furnishes the "childer" with many a savoury bite besides. In fact, if any food can, in these high-priced days, be called economic, bacon deserves the name, for it goes further than any other meat. My remarks, therefore, must be taken to apply only to those who have a choice, and who therefore should use it more as a relish than as the principal ingredient in the family bill of fare.

LESSON XII.

THE GIST OF THE WHOLE MATTER.

Now let us sum up what we have been trying to teach and to learn in this little book. To begin with, we will run through the first part, which is perhaps rather alarming on account of its hard words, and see what has been said.

No one will deny the importance of urging rich and poor alike, in the present state of things, to try and economize the fuel and food which they may have at their disposal. When I use the word economize, and apply it to rich people, I mean it to bear a wider significance than when I speak of the very poor, with whom it is an absolute necessity. It is just because there is not this absolute necessity on the score of expenditure, that a due attention to the principles of economy in food and fuel sits so gracefully on a rich person. I do not mean that only two fires should be lighted in a splendid mansion, or that its inmates should gather every day around a dinner of bone-soup or a lunch of bread and cheese. That would of course be absurd nonsense, and no one is so short-sighted as not to perceive that such economy would starve a good many thousand people in other grades of life. What I mean is, that in all households, beginning with those costly establishments where the duty devolves on a steward or housekeeper, there should be such arrangements, such training, such recognized principles, that the possibility of *waste* should be reduced to the lowest point. Everyone will acknowledge that in what are called "great kitchens," the "waste,"—the broken victuals, scraps, crusts, bones, and so forth—would feed many a poor and hungry family. All I say, then, is: "Let it feed such families: don't let it be thrown away, or sold as refuse.". When we have made the most of everything, there will still be quite enough refuse in the world, without adding to it portions of food which would be a boon and a blessing to a starving child. The same with fuel. Let people who can afford to pay for coals have as many fires as they choose, but let them take care that the coals are fairly used and made the most of, cinders and all, so will there be more left in the market for those to whom a hundredweight of coal is of more importance than is a ton to a rich man. Let such people have grates and stoves, and all the new inventions for the economy of fuel, and then, if everybody makes a conscience of being careful with their coals—economical without being stingy, but insisting on every cinder being duly used, or even given away, instead of finding its way into the dust-hole—we shall not perhaps have constant alarms of scarcity and famine prices.

So much can rich people do to help; but those in the lower grades of society can do a great deal more; and I am persuaded that the chief reason a great deal more is not done is because people don't know how to do it. The mistress of a middle-class household considers that she fulfils the whole duties of her position by giving a few languid orders to her servants, which they obey or not, according to their several dispositions. By all means let her confine herself to this feeble style of housekeeping until she knows *how* the things should be done, for until then it is better she should not interfere. If everything was exactly as it should be, if cooks knew not only how to lay and light fires, but to cook exquisitely, it would be very delightful, and we might all live happy ever after. But, unfortunately, we seem to be a long way from such a desirable state of things; and complaints of the bad, and an outcry for good, servants grow louder every year. Now, it appears to me that good mistresses are just as much needed as good servants, mistresses who are capable of explaining kindly and clearly to a servant how and why their duties—or such portion of their duties as they are ignorant of—should be performed. Explanation is a good deal better than scolding, and the practical knowledge from which such explanations should spring is quite compatible with the utmost refinement and cultivation of the mind. I don't want ladies to do the servants' work; I only want them to have the opportunity of learning to explain how such work should be performed, and to understand, even in theory, why and wherefore certain causes bring about certain results in domestic economy.

Let us take the mistress of an ordinary middle-class household, a household where the husband works hard to make an income of from 500*l*. to 1,000*l*. a year, on which four or five children have to be educated and set forth in the world, and perhaps relations to be helped besides (for poor people generally have to help their relations). Ten years ago it would have been, for that rank of life, almost a large income. Nowadays it is a very small one, and it has therefore become more than ever of grave importance that the person on whom its management chiefly depends should know something besides music and drawing. Well, then, this typical lady shall be amiable, intelligent, anxious to do her best for her family and household, and yet what state of things shall we be tolerably sure to find in such a house? In the nursery, " Missis " is all that is capable and useful. She thoroughly understands how to provide for the health and pretty toilettes of her nice little children. She and Nurse get on very well; they have a mutual respect and confidence in each other's " knowledgeableness," and a thorough belief in each other's capacity. All is right at the top of the house. On the next story the lady is not quite so certain of her ground. She has indeed slender theories on the subject of dust, and, we will hope, a wholesome love of fresh air, but a new housemaid will probably find that she can do pretty much as she likes in her own department.

But it is not till we come down to the kitchen that we begin to suspect there is a screw loose somewhere. *If* our lady has been fortunate enough to stumble upon a cook who for 14*l.* or 16*l.* a year will cook savoury meals for her every day of her life; a cook who is as clean as she is clever, and as honest as she is sober, then indeed there will be peace and harmony in that establishment, unless the cook should happen to have a bad temper. But how is it if the cook be merely an ignorant, honest, "willing" young woman? Who is to teach her? How and where is she to be trained? That has hitherto been the great difficulty of English middle-class life, and it is to remove, or at all events to give those who wish it an opportunity of removing it, that the National School of Cookery is to be established at South Kensington. Everything cannot be done in a moment; unsuspected needs will crop up, an extended sphere will necessitate wider arrangements; but I can safely affirm that the point which will be steadily kept in view by the Committee is this great need of the English people—the want of some place where a girl or woman can be taught how to cook. It is not necessary for ladies to bend over the fire and harden their palms with saucepan handles, for it is easier to teach an educated person by theory than an uneducated one; and a lady will carry away a great deal of useful knowledge from a lecture where a cook-maid would have been swamped by words and phrases above her capacity. There will therefore be both forms of education; but, so far as my own experience goes, and speaking confidentially, I should have been very thankful for both opportunities of practical instruction before I went to New Zealand. I might then perhaps have been saved many an anxious moment, to say nothing of constant culinary discomfiture. I *did* go down to a friend's kitchen more than once, and try what knowledge I could pick up, but I was so bewildered by the size and splendour of the *batterie-de-cuisine*, and the cook would persist in regarding my desire for information as either a whim or a joke on my part, so that it ended by my learning nothing whatever which proved of any practical use to me. To begin with, I could not explain to the cook what I wanted to know; I could not even say where my ignorance began or where it ended, though indeed I found out afterwards that it would have been well to have established some infallible test for ascertaining when the kettle boiled. What experiments even in this line were necessary when I set up for myself! including one recipe of turning the kitchen poker into a sort of tuning-fork, and holding the handle to my ear, whilst the poker-point rested on the lid of the kettle. That method soon fell into disfavour, for it used generally to result in upsetting the whole affair and extinguishing the kitchen fire.

Well, then, to return to the purpose of this slender volume. If it even awakens a sense of ignorance in its readers, something will have been gained, for I am much mistaken in my knowledge of women of my own class and position in life, as well as of those in a higher rank, if, when once they feel the need of practical instruction and improvement in their domestic arrangements, the next step will not be to endeavour to acquire that knowledge. Also, I hope and believe that the artisan's young wife, who feels the commissariat and cooking a heavy burthen on her mind and her hands, will set to work to learn how and why certain food-substances are more wholesome and therefore more economical than others, and in what fashion they should be cooked so as to make them go further and render them palatable.

Lower than this grade in our social scale it seems hard to go. It is too much to expect the crowds whose daily bread is a perpetual miracle, to have the time and the means to learn to cook better. When it is generally a matter of chance and locality what sort of food they can provide for themselves and their children, it seems a bitter mockery to tell them this, that, and the other is the most nourishing diet, or to recommend rump-steaks to them instead of bread and dripping. But here, those rich and benevolent people, whose comforts and luxuries have been and will be secured to themselves and their families for many a day, may possibly find another outlet for that spring of human sympathy and charity which—whatever pessimists may say to the contrary—runs bright and sparkling beneath our natures, and wells up to make many a green and blessed spot in our own lives and those of others.

Let us look for a moment at our country villages, and think how often it happens that the Squire's and the Rector's wife is asked to take some well-behaved cottage-girl and "learn" her to cook.

With the best will in world, what can these kind ladies do? With a sigh they will consent, and return home to announce—probably with some trepidation—to their cook, that "a new girl" is coming. This means a year of misery and discomfort to everybody. The cook does not care about teaching the girl, and will most likely take but slender pains to do so. The girl feels that she is only on sufferance in the kitchen, and is in a false position there, besides. It will probably be very difficult, if not impossible, for her to get anything like a regular useful lesson from her aggrieved instructress. Everything that is broken in the kitchen is laid to her charge, and at the end of the year I question whether, even under the most favourable circumstances, such a girl can possibly have learned anything which will be of real practical value to her. As soon as ever she begins to have a dawning idea on the subject of a mutton-chop, she must go elsewhere and make room for another beginner. Now, the same money which would keep this girl for a year, would give her proper instruction in a proper place.

How constantly it happens that a young woman who is happily placed as housemaid or nursemaid, or apprenticed to a trade, loses her mother, and it becomes absolutely necessary that she should give up her situation and return home to fill, as best she may, her mother's vacant place. Such a girl has probably never cooked a meal for herself in her life. She may return home with an earnest and affectionate desire to do her best for her father's and brothers' comfort, but can she know by inspiration how to cook their meals? Even in my own limited experience I have repeatedly heard laments on this score, and felt myself at the same time quite powerless to help beyond the vague suggestion that the beginner should ask Mrs. So-and-so to show her a little how to cook; Mrs. So-and-so knowing probably very little herself.

Many hundreds and thousands of people in London and our other cities and watering-places live, at all events for a certain portion of the year, in lodgings, or, as they are more elegantly styled, furnished apartments. Imagine a monster meeting of lodgers in the Albert Hall, assembled to proclaim their greatest grievance. Would there not be one universal roar of "The food"?

I have occasionally lived in lodgings myself, and I can speak from my own experience, feeling confident that it will represent the experience of a considerable portion of the houseless community. I found invariably civility, generally cleanliness (or at all events that is a remediable evil), and, with scarcely any exception, *vile food*. When I complained, the stereotyped answer, given in a very hopeless tone, used to be: "Well, ma'am, I know it's not exactly right, but it's the gal; you see, she don't know nothing, and I can't cook myself, not to say well." Now, why can't the "gal" cook, poor soul? Has she ever been taught, or had even a chance of learning? Do we put ever so willing a man to fire an Armstrong gun or set up type without the slightest previous instruction on the subject? Why should a "gal" be taken from her school life (this is imagining the most favourable conditions), and suddenly be expected to know how to cook, especially when her teacher is confessedly as ignorant as herself? The only bright exception to this rule is when a girl has had the rare good fortune to be trained in some charitable institution, where she has been properly taught to *cook* as well as to scrub and clean, and to keep herself neat and tidy, even whilst she is working. Yet, as I write the words "rare good fortune," a remorseful pang comes over me; for, however such training may benefit the poor child and her employers in after years, it has probably been necessary, in order for her to be admitted into such an institution, that she should have been a waif or stray, an orphan, or a poor deserted child, or exceptionally wretched in some way, and it is from her very homelessness and helplessness that what I find myself calling her "rare good fortune" has sprung.

I have already alluded in another place (page 36) to the case of the domestic servant who has been a housemaid or a nursemaid, or waited on ladies, and who perhaps marries and finds herself in a nice little home which it becomes her duty to keep bright and clean. She can do everything except cook, but I venture to say she will find this a great difficulty, and there will be a good deal of unconscious waste and extravagance before even the Rubicon of fried bacon is passed.

It would be a good opportunity for this class of servants to learn cooking at the National School when families go out of town for the autumn, and two or three servants are left in an empty house to while away a couple of months as best they can. I do not want to curtail or interfere with any one's holiday, but it could scarcely be a grievance to a young woman who is perhaps looking forward to a little home of her own some not very distant day, to have the opportunity of taking lessons in the art of cooking her husband's meals. Many of our subscribers may be fortunate enough to possess cooks who are masters or mistresses of their science, and to whom the word instruction dare not be mentioned. What I would venture to suggest to such people is, that although they may not need instruction for their cooks, they might utilize the advantages which their subscriptions will give them, for the benefit of their younger servants or even of their tenants' daughters.

The great point which I have reason to believe the Committee of the National School of Cookery will insist upon is, *thoroughness.* No one will be allowed to run, or try to run, before she can walk. The elementary knowledge of how to light and manage a kitchen fire, of scrupulous cleanliness in pots and pans, of attention to a thousand small but all-important details, will be taught and insisted upon before the learner is allowed to do anything worthy of the name of cooking. She will then probably be surprised to find how comparatively easy it will be to acquire the art, and she may be very sure she will not be allowed to try a second thing until she can do the first, if it be only boiling a kettle or toasting a piece of bread to perfection.

Such is the plan for complete beginners—who, by the way, generally prove the most successful pupils;—but for servants or artisans' wives who wish to "better" themselves in their kitchens, there will be a different mode of instruction, into which we need not enter here. Ladies will also have an opportunity either of sitting in a chair and listening to a lecture or series of lectures on cooking, beginning with a mutton-chop and ending with a *soufflé*, or they may turn back their sleeves, take off their rings and bracelets, and try for themselves. It will be hard if any eager inquirer does not find some course or class to meet her needs; and it is to be hoped that whatever excuse may hereafter be urged for our national bad cookery, the reproach of the want of a place and opportunity of instruction will be done away with for ever.

There is but one parting remark I have to make. It is this. The National School of Cookery is not a mercantile undertaking. I have no wish to attempt to throw discredit upon such undertakings, but simply to state the School of Cookery at South Kensington is not one. There will be no question of dividends or bonuses, nor will there be shareholders whose interests and pockets must be considered. The School has every reason to expect that it will be liberally supported by contributions and donations; if it finds itself mistaken in that expectation, it will close its doors, and there will be no harm done to anybody. It is managed by a Committee of gentlemen whose names are a sufficient guarantee for their actions, and no one of them will be individually a penny the richer or the poorer, whether the undertaking succeeds or not. If the School be well and liberally supported, it will be a sign that the need of improvement in cooking is felt by all classes, and for every shilling subscribed it is the intention of the Committee to afford means of instruction. The more money which is forthcoming, the more widely-spread will be the benefit which the promoters of the National School of Cookery hope and believe it is capable of producing.

THE END.

LONDON: RICHARD CLAY & SONS, PRINTERS,

www.ingramcontent.com/pod-product-compliance
Lightning Source LLC
Chambersburg PA
CBHW031410160426
43196CB00007B/968